P9-AGE-092

OXFORD ENGLISH MONOGRAPHS

General Editors

THE SAVAGE AND THE CITY

in the work of

T. S. ELIOT

ROBERT CRAWFORD

CLARENDON PRESS · OXFORD
1987

Oxford University Press, Walton Street, Oxford OX2 6DP
Oxford New York Toronto
Delhi Bombay Calcutta Madras Karachi
Petaling Jaya Singapore Hong Kong Tokyo
. Nairobi Dar es Salaam Cape Town
Melbourne Auckland
and associated companies in
Beirut Berlin Ibadan Nicosia

Oxford is a trade mark of Oxford University Press

Published in the United States
by Oxford University Press, New York

British Library Cataloguing in Publication Data
Crawford, Robert
The savage and the city in the work of
T. S. Eliot. — (Oxford English monographs).
1. Eliot, T. S. — Criticism and interpretation.
I. Title
828'.91209   PS3509.L43Z/
ISBN 0-19-812869-X

Library of Congress Cataloging in Publication Data
Crawford, Robert.
The savage and the city in the work of T. S. Eliot.
(Oxford English monographs)
Based on the author's thesis (D.Phil.) — Oxford University.
Bibliography: p. Includes index.
1. Eliot, T. S. (Thomas Stearns), 1888–1965 —
Criticism and interpretation. 2. Man, Primitive, in
literature. 3. Primitivism in literature. 4. Cities and
towns in literature. 5. City and town life in literature.
6. Literature and anthropology. I. Title. II. Series.
PS3509.L43Z654 1987 821'.912 87–7845
ISBN 0-19-812869-X (U.S.)

Phototypeset by Dobbie Typesetting Service,
Plymouth, Devon
Printed in Great Britain
at the University Printing House, Oxford
by David Stanford
Printer to the University

For my parents

ACKNOWLEDGEMENTS

I AM particularly grateful to Mrs T. S. Eliot for her helpful interest during the writing of this book, for her permission to examine restricted materials, and for allowing me to quote from unpublished writings by T. S. Eliot in the Berg Collection, Astor Lennox and Tilden Foundations, New York Public Library; Butler Library, Columbia University; the Hayward Bequest, King's College, Cambridge; Houghton Library and Pusey Library, Harvard University; the Huntington Library, California. All such quotations are reprinted by permission of Mrs Valerie Eliot and Faber & Faber, Ltd., ©Mrs Valerie Eliot, 1987. For their co-operation and courtesy I thank the staffs of these libraries, and of the other libraries detailed in the Bibliography. I am again grateful to Mrs Eliot and to Faber & Faber for permission to quote generously from numerous uncollected literary writings by T. S. Eliot, and for permission to quote from *The Waste Land: A Facsimile and Transcript* by T. S. Eliot edited by Valerie Eliot. Material is reprinted by permission of Faber & Faber, Ltd. from the following books by T. S. Eliot: *Collected Poems 1909-1962*, *Collected Plays*, *Old Possum's Book of Practical Cats*, *For Lancelot Andrewes*, *After Strange Gods*, *Selected Essays*, *On Poetry and Poets*, *Notes towards the Definition of Culture*, *The Use of Poetry and the Use of Criticism*, *Knowledge and Experience in the Philosophy of F. H. Bradley*, *To Criticize the Critic*, and *The Idea of a Christian Society*; material is reprinted by permission of Harcourt Brace Jovanovich, Inc., from T. S. Eliot's *Collected Poems 1909-1962*, *Four Quartets*, *Old Possum's Book of Practical Cats*, *Murder in the Cathedral*, *The Family Reunion*, *The Cocktail Party*, *Selected Essays*, *The Idea of a Christian Society*, and *Notes towards the Definition of Culture*; material is reprinted by permission of Harvard University Press from *The Use of Poetry and the Use of Criticism*; material is reproduced by permission of Farrar, Straus & Giroux, Inc., from Eliot's *Poems Written in Early Youth*, *On Poetry and Poets*, *Knowledge and Experience in the Philosophy of F. H. Bradley*, and *To Criticize the Critic*.

Special thanks are due to Professor Richard Ellmann for the generosity and humanity with which he supervised the Oxford

D.Phil. thesis which formed the first version of this book. To Richard and Mary Ellmann I owe various delightful debts. The writing of the thesis at Balliol College, Oxford, was made possible by the generous award of a Postgraduate Scholarship by the Carnegie Trust for the Universities of Scotland along with a Snell Exhibition and Newlands Scholarship given by the University of Glasgow and Balliol College, as well as other grants given by these institutions, and by the University of Oxford Graduate Studies Committee. An additional Carnegie Grant made possible a very valuable research visit to New York, Yale, and Harvard. The Keeper and Trustees of the Matthiessen Room, Eliot House, Harvard University, were most accommodating to me during my stay as visiting scholar there. The thesis was completed during my first term as Elizabeth Wordsworth Junior Research Fellow at St Hugh's College, Oxford, and subsequently revised there for book publication. I am grateful to the Principal and Fellows of St Hugh's College for electing me to this post, and for making substantial grants towards the payment of copyright expenses incurred in the production of this book.

In addition to the general and large debt I owe to my teachers and colleagues in Glasgow and Oxford I am grateful for specific help afforded me by the following individuals, either in conversation or in correspondence: Mr Peter Ackroyd, Professor William Alfred, Dr Nick Allen, Professor William Arrowsmith, Professor W. J. Bate, Dr Jewel Spears Brooker, Dr P. Bulloch, Professor John Carey, Mr Malcolm Chapman, Professor John Creaser, Miss P. Deery, Professor John Finlay, Dr Donald Gallup, the late Dame Helen Gardner, Dr Lyndall Gordon, Dr Piers Gray, Dr Michael Halls, Professor Alan Heimert, Dr Roger Highfield, Professor Harry Levin, Dr Godfrey Lienhardt, Professor Walton Litz, Professor Edwin Morgan, Professor Jeffrey Perl, Mr Jack Rillie, Mr Kayode Robbin-Coker, Mr Carl Schmidt, Professor Ronald Schuchard, Miss E. M. Sillitto, Mr Paul Turner, Dr Anne Varty, Dr T. H. H. Vuong, Dr Anthony R. Walker, Dr George Abbott White.

Versions of some sections of this book have appeared in *Essays in Criticism*, the *Journal of the Anthropological Society of Oxford*, and *Victorian Poetry*. Final thanks must go to my co-editors of *Verse*, Dr David Kinloch and Professor Henry Hart, for patiently collaborating with me in editing the magazine while this book was being written.

CONTENTS

LIST OF FIGURES

ABBREVIATIONS

IN the notes all books or articles are fully detailed when first mentioned and thereafter denoted by the author's name. In the case of works repeatedly cited, these may be found in the Bibliography, sections 4 and 5. Where several works by an author are concerned, each is identified by date of publication, e.g. Pound (1950). Where no author is mentioned, the writer is T. S. Eliot, unless it is stated that the piece is an unsigned one. For convenience, where Eliot articles are being mentioned in abbreviated form, these are followed by the appropriate reference number in Gallup's Bibliography, where such a reference number exists, e.g. 'A Prediction' (C153). *Nation* throughout refers to the New York *Nation*. In addition, the following abbreviations are used for works by Eliot:

ASG	*After Strange Gods: A Primer of Modern Heresy* (London: Faber & Faber, 1934)
CC	*To Criticize the Critic and Other Writings* (London: Faber & Faber, 1965)
CPP	*The Complete Poems and Plays of T. S. Eliot* (London: Faber & Faber, 1969)
Facsimile	*The Waste Land: A Facsimile and Transcript* ed. Valerie Eliot (London: Faber & Faber, 1971)
FLA	*For Lancelot Andrewes* (1928; reissued London: Faber & Faber, 1970)
ICS	*The Idea of a Christian Society and Other Writings*, Second Edition (London: Faber & Faber, 1982)
IPR	Essay on the Interpretation of Primitive Ritual (1913) now in King's College Library.
KE	*Knowledge and Experience in the Philosophy of F. H. Bradley* (London: Faber & Faber, 1964)
Notes	*Notes Towards the Definition of Culture*, Second (1962) Edition (London: Faber & Faber, 1962)
OPP	*On Poetry and Poets* (London: Faber & Faber, 1957)
Rock	*The Rock: A Pageant Play* (London: Faber & Faber, 1934)
SE	*Selected Essays*, Third Enlarged Edition (London: Faber & Faber, 1951)

SW   *The Sacred Wood*, Second (1928) Edition
            (London: Methuen, 1928)
Use   *The Use of Poetry and the Use of Criticism*,
            Second (1964) Edition (London: Faber & Faber,
            1964)

INTRODUCTION

THE most striking aspect of the work of T. S. Eliot is its constancy. From his early years Eliot displayed a sensibility fascinated by the bringing together of apparent opposites. That sensibility flowered in the union of tradition and modernity, of Europe and America, of blasphemy and religion, of slapstick mischief and poker-faced enthronement. Most excitingly, and perhaps most revealingly, it flowered throughout his greatest work in the uniting of the world of the savage with the world of the city.

At the age of ten, Eliot celebrated twice in a childhood magazine the betrothal of Miss End and Mr Front; clearly he was amused by this meeting of opposites, and the change of name which would ensue.[1] In 'Portrait of a Lady', it is the lady, whose affected and empty refinement is mocked, who makes the aside

(But our beginnings never know our ends!)

For Eliot, beginnings and ends were always closely bound together, though knowledge of their connecting pattern might emerge only much later. As he put it in 'Little Gidding',

What we call the beginning is often the end
And to make an end is to make a beginning.

This preoccupation with bringing together apparent contraries lasted throughout Eliot's life. The *Fireside*, that little magazine which he edited in St Louis during early 1899, also records in its customary pencil the elopement of Mr Up and Miss Down.[2] In 1912 at Harvard, James Woods would point Eliot to Diels's edition of Heraclitus whose philosophy saw the world as the result of contrary strains, and over twenty years after that, Eliot would select as an epigraph for his *Four Quartets* a fragment from his own copy of Diels's edition: ὁδὸς ἄνω κάτω μία καὶ ὡυτή. 'The way up and the way down are one.'[3]

---

[1] *Fireside* (unpublished childhood magazine), 28 Jan. 1899, p. [3], and 2 Jan. 1899, p. [5] (Houghton).

[2] Ibid. 2 Jan. 1899, p. [8].

[3] Notes on J. H. Woods's Course, Philosophy 12, 1911–12, [16–17] (Houghton); CPP, p. 171.

2 *Introduction*

'When a poet's mind is perfectly equipped for its work,' Eliot wrote in 'The Metaphysical Poets' (1921), 'it is constantly amalgamating disparate experience.'[4] Clearly this concept was developed out of Eliot's enthusiasm for the *discordia concors* of Metaphysical poetry and the strange juxtapositions of later nineteenth-century French verse, but Eliot himself was attracted continually to setting explicit contraries together, Prufrocks against prophets. So, the repellently primitive confronts the brilliantly sophisticated when, in 'Burbank with a Baedeker: Bleistein with a Cigar', a subhuman eye (or the eye of someone suffering from Graves' disease) confronts the sophisticated optics of a perspective of Canaletto. 'Gerontion' sets withered inanity against full-blooded heroism. The placing of eastern beside western asceticism at the end of the third section of *The Waste Land*, Eliot wrote wryly in his notes, 'is not an accident'.[5] With regard to 'Marina', he explained in a letter to Sir Michael Sadler, he intended a mixture of Seneca's Hercules awaking to discover that he has slaughtered his own children and Shakespeare's Pericles who, excitedly half-asleep, discovers his living lost daughter.[6] Meetings of polar opposites are vital to the pattern of Eliot's work. One of the central interests of the *Four Quartets* concerns

> The point of intersection of the timeless
> With time . . .

'Little Gidding' ends with a resolution of themes developed in earlier movements, when 'the fire and the rose are one'. This last quartet concludes with a return to a struggled-for, childlike simplicity, pronouncing that

> the end of all our exploring
> Will be to arrive where we started
> And know the place for the first time.

The knot of ends and beginnings is once again being tied.

This study examines one particular combination of apparent opposites in Eliot's work, his linking of the most primitive and barbaric with the most sophisticatedly urban. His study of primitive societies went hand in hand with his study of decadent ones. This

---

[4] *SE*, p. 287. [5] *CPP*, p. 79.
[6] Postscript from Eliot's 9 May 1930 letter to Sir Michael Sadler (Bodleian), quoted in A. D. Moody, *Thomas Stearns Eliot, Poet* (Cambridge: Cambridge University Press, 1979; first paperback edn. 1980), p. 357 n. 33.

key preoccupation with savage and city emanates from his own childhood and is strengthened by his work at Harvard and in London,
before it emerges as crucial to some of his greatest poetry in *The
Waste Land*, 'The Hollow Men', and *Sweeney Agonistes*. The need
to sharpen this theme continues in the later work. The *Quartets*
modify the earlier treatment, yet continue to provide not only the
urban world of London, but also

> the backward half-look
> Over the shoulder, towards the primitive terror.

Such a primitive terror had already been seen by the West End
audiences of *The Family Reunion*, confronted by the Furies, while
The Cocktail Party shocked its audiences with its sophisticated
town gathering interrupted by news of a crucifixion among jungle
tribesmen. Though this book concentrates on the major poems, the
combination of savage and city runs through Eliot's career. It is
present in his scurrilous unpublished King Bolo poems, whose
eponymous 'hero' is a savage in a bowler hat, as well as in Eliot's
decidedly more sober social criticism where the legacy of earlier
anthropologists moulds his view of modern society. The term 'savage'
is employed here, because that is the term used by most of the
anthropologists and anthropologically influenced scholars whose
works Eliot read. Often he followed them in his use of the word.
The 'city' is not simply London with its metropolitan sophisticate
and its City financial district, but also the drab urban landscape
which Eliot annexed as his poetic territory from his first book
onwards. Eliot's urban landscape is a strange mixture of lived and
literary experience. Often its nourishment came from unexpected
ground. Eliot in 1932 wrote that 'some of Dickens' novels stand
for London', and that he was particularly moved by 'the Chancery
Prisoner in *Pickwick*, of whom is said finally, *He has got his discharge,
by God*'.[7] *The Waste Land* is certainly about prisoners, and involves
'different voices', but essentially the poem's city owes more to Kipling
and Victorian poetry, both of which were childhood favourites.[8]
Similarly, an interest in the savage seems to date from Eliot's early
childhood. Though the two elements were linked in the Victorian
phrase 'city savage', Eliot's linking of them produced startling results.

---

[7] 'Preface' to Charles-Louis Philippe, *Bubu of Montparnasse* (trans. L. Vail; Paris:
Crosby Continental Editions, 1932), pp. viii and xi.
[8] *Facsimile*, p. 5, and p. 125 n. 1.

Understanding this peculiar combination in his work makes possible a new interpretation of *The Waste Land* and lets us see much more clearly what he was about in *Sweeney Agonistes*. It helps restore to the reading of Eliot a sense of the excitement, the openness and breadth of his art, an art which drew on the *intichiuma* rites of the Australian aborigines as well as on the rites of the Christian Church; which looked to jungles, jazz, and the Wild West as well as to the tasks of bank clerk and churchwarden. Eliot's poetry is one of struggle, longing, and mischief, as multifaceted as it is intense. The study of the savage and the city in his work opens doors in his poetry which lead to predictable and less predictable destinations, including Mendelian heredity, evolution, blood-and-thunder stories, Melanesia, and Lloyds Bank. Most important is the realization that an understanding of Eliot's continuing concerns with savage, city, and their connection helps us appreciate not only the ferocity and liveliness of Old Possum, but also the way in which texts as diverse as 'Growltiger's Last Stand', 'The Dry Salvages', and *The Waste Land* emerge from the same temperament, showing lighter and darker sides of an apocalyptic imagination. In what follows, one aspect of Eliot's work has not been stressed: its humour. Perhaps the recent outstanding success of the musical *Cats* renders such a stress unnecessary. Eliot the vaudeville fan and eulogist of Marie Lloyd would have relished that success. What is more the costume designer for the musical seems to have realized, more than most critics, that the imagination which produced Macavity and the 'Cat about Town' was the same imagination which, when young, delighted in and smiled at tales of adventure in the Wild West, which re-viewed earlier literature through anthropological eyes and which, when old, was thrilled by the opening of Tutuola's *My Life in the Bush of Ghosts*.[9] What are scrutinized in this book are the underpinnings of Eliot's writing.

[9] See Bernth Lindfors, 'Amos Tutuola's Search for a Publisher', *Journal of Commonwealth Literature*, vol. 17, no. 1 (1982), p. 103.

I

THE BEGINNING OF THE WILD WEST

the communication
Of the dead is tongued with fire beyond the language of the living

THE dead always spoke strongly to T. S. Eliot. Continually he heard their voices and saw their patterns of action reincarnated in the speech and behaviour of contemporary life. Often in his greatest work the primitive and forgotten burn ineradicably through the sophisticated, urban world. Eliot's own birthplace, St Louis, contained traces of a primitive past, and in that city Eliot was haunted by various voices, some from the pages of books, some from the stories of his own ancestors, which reached him 'tongued with fire'. He associated his grandfather, in particular, with such fiery communication, when he remembered that in his childhood the family's moral decisions were taken as if, Moses-like, his grandfather had brought down the tables of the Law.[1]

Eliot's grandfather died in 1887, a year before the poet's birth, but his influence continued. The conditions of Eliot's life were moulded pre-natally, grandfather's end becoming grandson's beginning. Eliot was an inheritor, born into a household which laid great stress on the inheritance of cultural codes, most notably his grandfather's law of social obligation.

This original Law of Public Service operated especially in three areas: the Church, the City, and the University. The Church meant, for us, the Unitarian Church of the Messiah, then situated in Locust Street, a few blocks west of my father's house and my grandmother's house; the City was St. Louis . . . the University was Washington University, then housed in a modest building in lower Washington Avenue.[2]

Eliot notably gives each institution a precise geographical location. All are contained in St Louis, as if that city were the boundary of his childhood. Certainly he liked to remember himself as a town child. He once described his 'urban habits' as 'pre-natal', and liked

---

[1] CC, p. 44.   [2] Ibid.

to remember not only how he had inherited, but also how for much of the year he was bound by, an urban landscape.[3] Eliot recalled living in a run-down, almost slummy area of the town after all the family's friends had moved to districts further west. But this was Eliot family ground, and they stayed on there, seldom moving beyond the town in these pre-automobile days. So, he insisted, his familiar landscape for nine months of each childhood year 'was almost exclusively urban', much of it being seedy and drab.[4] Under all the later concretions, Eliot maintained, the essential city of his poetry was St Louis.[5] Yet almost from the first, what he experienced was linked with and modified by other men's words, a heritage as inescapable as that of the city itself.

The first portrait of T. S. Eliot was painted by his sister Charlotte. The sitter was ten years old, and was painted as he read a book. That book no doubt kept the sitter still, but it is likely that the artist included it in her picture because it was a prop which brought out something of her subject's personality. After all, Charlotte had lived in the same house as her brother for ten years and must have known his character at least as well as any of Eliot's later iconographers. That her choice of prop was a good one is plain not only from the small boy's later careers, but also from a photograph taken about two years earlier. Aged eight, again Tom is on a wooden chair, but this time he is less formally posed, one leg folded beneath him, one foot on the ground as he hunches over his book, curled round it, possum-like, seemingly impervious to the shutter's click.[6]

When Eliot wrote of himself as 'a child of ten', he presented himself not in the posture selected by his sister, but as 'a small boy peering through sea-water in a rock-pool, and finding a sea-anemone for the first time'.[7] The context is a passage discussing the generation of poetic imagery and the way in which experiences may 'lie dormant' for long spaces of time before resurfacing in verse. But in his own verse there is a passage seeming to deal with the young reading Possum in a typical attitude when, in 'Animula',

---

[3] 'A Commentary', *Criterion*, Apr. 1938, p. 482.
[4] 'The Influence of Landscape upon the Poet', *Daedalus*, Spring 1960, pp. 421–2.
[5] Ibid., p. 422.
[6] T. S. Matthews, *Great Tom* (London: Weidenfeld and Nicolson, 1974), plates between pp. 108 and 109.
[7] *Use*, pp. 78–9.

> The pain of living and the drug of dreams
> Curl up the small soul in the window seat
> Behind the *Encyclopaedia Britannica.*

Literature for Eliot was always very much the possum's business, the text being associated with the curling up of the intimate self, hiding away beneath a cold or even apparently rebarbative exterior, 'the man who suffers' rolled up inside and hidden by the 'mind which creates'.[8] Eliot was already the possum before Pound nicknamed him. Around some books and phrases he curled himself so tightly that they became a part of him, part both of the public and of the secret self. As intermediaries, these texts formed go-betweens between the life of the sufferer and the public life of the creator. As other men's words, phrases such as Dante's '*Poi s'ascose nel foco che gli affina*' or Thomson's 'memory and desire' were public property. To use them was one of the most direct ways of being 'impersonal', yet, at the same time, other men's words might have the most intense personal significance. '*Fourmillante cité, cité pleine de rêves, / Où le spectre en plein jour raccroche le passant . . .* I knew what *that* meant,' Eliot declared, 'because I had lived it before I knew that I wanted to turn it into verse on my own account.'[9]

Eliot lived such a life only later. Though he recalled 'The pain of living and the drug of dreams', he more usually recalled childhood as a period of happiness, a happiness lost in later years, then finally rediscovered. Writing to Pound in November 1964, Eliot said that his second marriage brought him the first happiness he had known since childhood.[10] In Eliot's poetics of resuscitation, childhood happiness and pleasures associated with it continually resurface along with fleeting moments of illumination. So often it is the childhood world which underlies all later experience.

The secret self might be presented behind public words, occasionally breaking through them and, more commonly, adapting them to new ends, and the expression of that secret self was a perennial concern with Eliot. 'Prufrock' was written when Eliot was closely interested in the work of Bergson who was investigating, amongst other things, action, the will, and the way in which 'these two acts, perception and recollection, always interpenetrate each other'.[11] But for all

---

[8] *SE*, p. 18.　　　　　　　　　　　　　　　　　　　　　[9] *CC*, p. 127.
[10] Letter to Pound, 11 Nov. 1964 (Beinecke).
[11] Bergson, *Matter and Memory* (1896; trans., London: Swan Sonnenschein, 1911), p. 72.

Bergson's concern with the nervous system, when Prufrock, like so many of Eliot's later personae, comes to worry about self-expression, it is hard not to see him as endowed with Eliot's own recollection of a striking image seen in a St Louis newspaper when Eliot was eight.

It is impossible to say just what I mean!
But as if a magic lantern threw the nerves in patterns on a screen: . . .

In January 1897 the *St. Louis Daily Globe-Democrat* (a paper bought by Eliot's father and remembered by Eliot as the 'chief Republican organ' of St Louis) printed beside its unusually large and striking picture of this scene an article on 'Seeing the Brain', which points out how X-rays mean 'literally having one's thoughts read', and details what an uncomfortable process this is.[12] Eliot was always attracted to the secret word, whether 'The Word without a word, the Word within' of 'Ash-Wednesday' or the 'ineffable effable / Effanineffable / Deep and inscrutable singular Name' of 'The Naming of Cats'. At the same time, however, he feared the pain of self-revelation, of being exposed, and sought always a precedent, as if the previous use of an image or a phrase gave it a literary sanctification, consecrating it for his use. Frequently what appears original, fresh speech in Eliot turns out to be the language of the dead, revivified for his own ends. The near past continually resuscitates the distant. This makes his earliest reading particularly important, since it established the first precedents, sanctifying the earliest material. Similarly, the pattern of his childhood is important not only for the images of happiness which it gave him, but also because it became most profoundly embedded in all his later reading. Eliot's childhood is at the essence of his work.

However, because of the nature of that embedding, Eliot's childhood is most deeply hidden under the impersonal surface which he made his own theory of poetry. His childhood is as deeply buried as it is constantly revived. Metaphorically, at least, in this poetry, corpses are always sprouting. In the April 1929 number of the *Criterion*, Eliot remembered some of his childhood reading, including

[12] 'Seeing the Brain' (unsigned article), *St. Louis Daily Globe-Democrat*, Sunday Morning, 17 Jan. 1897, p. 39; Frank Morley, 'A Few Recollections of Eliot', in Allen Tate, ed., *T. S. Eliot: The Man and his Work* (London: Chatto and Windus, 1967), p. 109. Eliot remembered the paper in 'President Wilson', *New Statesman*, 12 May 1917, p. 140.

Fig. 1. ' as if a magic lantern threw the nerves in patterns on a screen'

the work of Conan Doyle.[13] Sherlock Holmes was to be a favourite with Eliot all his life. In 'The Adventure of the Copper Beeches', Watson complains to the reader about Holmes's apparent egotism after the great detective has criticized Watson for 'attempting to put colour and life into each of your statements' rather than being more plainly factual and relying on 'severe reasoning'. Holmes's retort is that 'If I claim full justice for my art, it is because it is an impersonal thing—a thing beyond myself.'[14] Even as a child, Eliot was attracted to this impersonal method, and much later he would explicitly connect Holmes's method with that of Paul Valéry, and, by implication, that of T. S. Eliot, when he wrote of Valéry's ideal poet who 'is to carry out the role of scientist as studiously as Sherlock Holmes did'.[15] Eliot himself, though he attempted to replace in 'Tradition and the Individual Talent' Pater's notions of the alchemy of art with scientific analogues involving filiated platinum, did not quite succeed as a chemist.[16] Nevertheless, he donned with some glee the garb of the 'unpoetic' banker poet, the proper City gent using in his early poetry many sordid urban landscapes, presented in sharp detail, in contrast to both the hazy lights of 1890s London and the traditional Romantic countryside. Holmes, for Eliot, was not only the scientist but the urban man.

> Sherlock Holmes reminds us always of the pleasant externals of nineteenth-century London. I believe he may continue to do so even for those who cannot remember the nineteenth century; though I cannot imagine what it would be like to read him for the first time in this volume, without the old illustrations.[17]

The old illustrations of Sidney Paget hardly ever show the streets of London, though they consistently present figures posed melodramatically in urban interiors. They are in keeping with the way violence breaks through the often clerkly world of the stories as when, in 'The Stockbroker's Clerk', Watson remembers how the title

[13] 'Sherlock Holmes and his Times', *Criterion*, Apr. 1929, pp. 552–6.
[14] A. Conan Doyle, 'The Adventure of the Copper Beeches', *The Adventures of Sherlock Holmes* (London: George Newnes, 1892), pp. 289–90. This edition contains the 'old illustrations' by Sidney Paget.
[15] 'Introduction' to Paul Valéry, *The Art of Poetry*, trans. Denise Folliot (London: Routledge & Kegan Paul, 1958), pp. xix–xx.
[16] See Grover Smith, *The Waste Land* (London: George Allen & Unwin, 1983), p. 23.
[17] 'Sherlock Holmes' (C283), p. 553.

character 'looked up at us' and 'it seemed to me that I had never looked upon a face which bore such marks of grief, and of something beyond grief—of a horror such as comes to few men in a lifetime'.[18] In Paget's illustration, as in Conan Doyle's text, we are poised on the brink of *Sweeney Agonistes*. Such horror would possess Eliot.

Holmes's urbanity and his urban world, though, are of clearer use to Eliot elsewhere. The passage where Holmes describes his 'art' as 'impersonal' goes on to describe the warm cheeriness of Holmes's room, while outside, 'A thick fog rolled down between the lines of dun-coloured houses, and the opposing windows loomed like dark, shapeless blurs through the heavy yellow wreaths.'[19] Eliot recalled that the 'yellow fog' of Prufrock, weighing heavily on that character as it sleepily paralysed him like ether, was drawn from the fog of St Louis factory chimneys.[20] But even in his childhood, this fog was seeping through the pages of Conan Doyle, just as later it would blend with the 'brouillard sale' of Baudelaire's 'Les Sept Vieillards'. Even before Eliot went to London, that city and St Louis were beginning to blend.

Holmes's London is often set against the natural world which invades it with a violence from beyond civilization.

It was in the latter days of September, and the equinoctial gales had set in with exceptional violence. All day the wind had screamed and the rain had beaten against the windows, so that even here in the heart of great, handmade London we were forced to raise our minds for the instant from the routine of life, and to recognize the presence of those great elemental forces which shriek at mankind through the bars of his civilization, like untamed beasts in a cage.[21]

Eliot's first visit to London, apparently in 1910 or 1911, would produce an unpublished poem 'Interlude in London' which juxtaposed the coming of spring with the unconcerned city.[22] In most of his early poetry impulses of nature are stifled by the moribund

---

[18] A. Conan Doyle, 'The Stockbroker's Clerk', *The Memoirs of Sherlock Holmes* (London: George Newnes, 1894), p. 67.

[19] 'The Adventure of the Copper Beeches', p. 290.

[20] Undated letter to Patricia Holmes (British Library).

[21] A. Conan Doyle, 'The Adventure of the Five Orange Pips', *The Adventures of Sherlock Holmes*, p. 108.

[22] In October 1910 Eliot bought his London *Baedeker* (King's College Library); 'Interlude in London' is dated April 1911 in the MS 'Complete Poems' (Berg).

composure of city life: 'evening quickens faintly in the street', only
to be extinguished by the ubiquitous *Boston Evening Transcript*. But
in *The Waste Land* the city and its whole civilization were to be
confronted with other, vastly elemental forces. Such a confrontation
was a commonplace of Eliot's childhood, something more familiar,
even, than the pages of Sherlock Holmes.

Growing up in St Louis beside the Mississippi, the 'big river', was to
Eliot an experience which he felt 'incommunicable' to those who had
not shared it.[23] Yet, as often, what seemed most 'incommunicable'
was what most fascinated him. 'Missouri and the Mississippi', he
wrote in 1930, 'have made a deeper impression on me than any other
part of the world.'[24] One of the features which impressed him most
was the annual spring freshet, when the river burst its banks and
inundated the surrounding land. 'It was a great treat to be taken
down to the Eads Bridge in flood time.'[25] In the spring of 1897,
for example, floods continually menaced St Louis, and by April
the river had reached unprecedented levels resulting in loss of life
and destruction of homes. It also washed away railway tracks.[26]
Throughout that cruel month, headlines in the *St Louis Globe-
Democrat* pronounced 'Flood Without Equal' for several weeks.[27]
Many years later Eliot remembered how at least twice at St Louis,
the western and eastern shores had been separated by a fall of bridges,
until the designer of the Eads Bridge found a structure to resist the
floods. In Eliot's childhood, the spring freshet often interrupted
railway travel, forcing the traveller to the east to take a steamboat
to Alton, at a higher level on the Illinois shore, before he could board
the train. 'The river is never wholly chartable; it changes its pace,
it shifts its channel, unaccountably; it may suddenly efface a sandbar,
and throw up another bar where before was navigable water.'[28]
Such a passage, of course, recalls the opening of 'The Dry Salvages'
where Eliot returns to that Mississippi of his childhood, a river
become 'only a problem confronting the builder of bridges'. But, as
a 'brown god' remaining a 'destroyer, reminder / Of what men

---

[23] 1930 sketch of his childhood, quoted in Moody, p. 4.
[24] Ibid.   [25] Ibid.
[26] See series on flooding, *St. Louis Globe-Democrat*, Jan.–Apr. 1897. Scenes of
railroad tracks washed away appear in issue of 2 Apr. 1897, p. 12.
[27] Headline, *St. Louis Globe-Democrat*, 10 Apr. 1897, front page.
[28] 'Introduction' to Samuel L. Clemens, *Huckleberry Finn* (London: The Cresset
Press, 1950), p. xiii.

choose to forget', it also presents something more primitive. The river, here, is a brown god because eagerly viewed in part through Mark Twain's eyes.[29] We may reserve this for discussion when considering *Four Quartets*. For the moment, it is important to remember how strong was Eliot's own experience of the 'big river' and its floods. Even as a boy, he knew the cruelty of the spring freshet elementally disturbing the life of the city, bringing revival, but also fatally dangerous. Later, when he read of primitive myths of resurrection involving overflowing waters, these would clothe an experience with which he had been intimately familiar as a child.

Eliot's parents did not allow their young son to read *Huckleberry Finn*, which they thought improper, with the result that he read it only in later life.[30] But there are other reasons, dating from Eliot's boyhood, why the river should be associated with primitive behaviour and why, when considering the landscape of his youth, he should see it in terms of 'gods' rather than simply God. For if Eliot was heir to a conventionally civilized heritage which, through the lineage of his family and its conduct, made his 'urban habits . . . pre-natal', he was also heir to glimpses of a more primitive life. Eliot's grandmother liked to recall how an Indian had sneaked into her kitchen and stolen a red ribbon from her hair.[31] As a child, Eliot's father had gone to Indian camping grounds and had been chased by Indians for imitating war whoops.[32] The Wild West was almost at its end in Eliot's childhood, but it was well remembered. It had become strangely entangled with the paraphernalia of urban society. Eliot remembered how in his early years 'the City was St. Louis— the utmost outskirts of which touched on Forest Park, terminus of the Olive Street streetcars, and to me, as a child, the beginning of the Wild West'.[33]

His genealogy and the stories he must have heard as a child are two reasons for Eliot's locating the Wild West's beginning just outside the city of his birth, but there is a more precise reason why it should have begun at Forest Park. Near the centre of Forest Park was a series of those prehistoric Indian mounds which are scattered in groups

[29] Ibid., p. xv.   [30] Ibid., p. vii.
[31] H. W. Eliot (Sr.), 'A Brief Autobiography' (MS in Washington University Archives, St Louis), p. 2.
[32] Peter Ackroyd, *T. S. Eliot* (London: Hamish Hamilton, 1984), p. 19.
[33] CC, p. 44.

throughout the alluvial plain of the Mississippi in the area known as the American Bottom and the largest of which is the Cahokia Group, on the east of the Mississippi and about twelve miles from St Louis. In Eliot's youth the Cahokia Mound was famed as the largest terraced earthwork in the United States. In 1914, celebrating the city's one hundred and fiftieth anniversary, the inhabitants gathered in Forest Park to see a pageant including representation of 'the aboriginal life of the Mississippi Valley', and later a performance of *Saint Louis: A Civic Masque* in which Cahokia, the Mound-Builder featured.[34] But the citizens of St Louis had for some time been very proud of their city, and the origin of the mounds had been a subject of discussion in Eliot's boyhood. Their physical presence, making the savage past a resolute part of the urban present, was still felt. When Eliot was six, on the less fashionable eastern side of St Louis where he and his family lived, a mound was destroyed which contained many human bones. Eliot mentions an earthen mound in a nonsense rhyme of 1899.[35] In 1904 D. I. Bushnell published results of excavations carried out in the autumn of 1901, when Eliot was thirteen, at the seven mounds in Forest Park, concluding that they were the sites of Indian lodges.[36] The presence of such remains would seem a good reason for Eliot's Wild West beginning specifically at Forest Park on the western, genteel side of town, and it demonstrates how physically, in the poet's earliest environment, city and savage were inextricably bound together.

In St Louis, Forest Park was also the site of the town zoo. As a child Eliot went there and photographed a buffalo, remembering his disappointment because it was more mangy looking than it should have been.[37] This buffalo, staple diet of the story-book Red Indians, is another reason for Forest Park being the beginning of the boy's Wild West. But Eliot's visit there was an act sanctified, as so many of his poetic acts were to be sanctified, by his reading.

The eleven-year-old Eliot was clearly attracted to blood-and-thunder stories, being himself the author of several, including

---

[34] Herbert Howarth, *Notes on Some Figures Behind T. S. Eliot* (London: Chatto and Windus, 1965), p. 47. Though Eliot was not in St Louis at this time, he appears to have known of the pageant, since he remarks on Civil Pageants in the context of American Universities, in a letter to Pound, Apr. 1915 (Beinecke).

[35] 'Poet's Corner', *Fireside*, no. 6, 2 Jan. 1899 (Houghton).

[36] D. I. Bushnell, jun., *The Cahokia and Surrounding Mound Groups* (Cambridge, Mass: Peabody Museum, 1904).

[37] Matthews, p. 13, using Eliot's 1930 reminiscence (see n. 23 above).

'Pony Jim', whose hero crosses plains and confronts Indians, 'Up
the Amazon' (more Indians), and 'Rattlesnake Bob', subtitled 'A Boy
at Santiago'.[38] His pencilled childhood 'magazine', the *Fireside*,
advertises a made-up book as 'a bloodcurdling tale'.[39] Forming an
apparent contrast with this is *Eliot's Floral Magazine*, only one issue
of which survives (February 1899), a more peaceable production,
cataloguing flowers with long names, though also mocking poetical
effusions about them. One flower comes from the South Seas.[40]

Such an odd combination of botany and bloodthirstiness seems less
odd when we realize that it represents an early literary enthusiasm.
Though she succeeded in preventing him from reading *Huckleberry
Finn*, Thomas's mother became anxious about his addiction to
another popular author, in whose works he had read about the Wild
West, bloodcurdling adventures at sea, and the dark deeds of savages
in various far-flung parts. Mrs Eliot tried to persuade her younger
son to read Macaulay's history of England instead. Fortunately, she
failed.[41]

This author who so disturbed Eliot's mother was Captain Mayne
Reid, probably the most popular American children's writer of the
day. His most famous book, *The Boy Hunters*, subtitled *Adventures
in Search of a White Buffalo*, was reprinted in America at least
twenty-one times between 1853, its first publication, and 1900, and
was for a time included in Everyman's Library as *The Boy Hunters
of the Mississippi*. Eliot himself never named the individual titles,
but, as will be obvious, Mayne Reid's novels run to a pattern. So
I shall concentrate on *The Boy Hunters*, but shall bring in a few of
the most popular of the other works as well.

In some respects it is surprising that Eliot's mother disapproved
of the attention which her son was devoting to Mayne Reid's novels.
Their intentions seem most suitably educational, acceptable, surely,
to the family of the founder of Washington University. The dedication
of *The Boy Hunters* certainly strives to achieve a lofty tone, being
addressed to boy readers in England and America (in block capitals)

[38] 'Pony Jim', by Dimey Novles (*sic*, sc. T. S. Eliot), *Fireside*, no. 8, 30 Jan.
1899; 'Up the Amazon', by Ferdino Peleigs (*sic*, sc. T. S. Eliot), *Fireside*, no. 11,
11 Feb. 1899; 'Rattlesnake Bob: A Boy at Santiago', by Hughly Stories (*sic*, sc.
T. S. Eliot), *Fireside*, nos. 4–5, 2 Jan. 1899 (Houghton).
[39] 'New Books', *Fireside*, no. 11, 11 Feb. 1899 (Houghton).
[40] *Eliot's Floral Magazine* (Houghton).
[41] 'Prize-Day Address at the Methodist Girls School at Penzance' (King's College
Library).

THAT IT MAY IMPRESS THEM,
SO AS TO CREATE A TASTE FOR THAT MOST REFINING STUDY,
THE STUDY OF NATURE—

THAT IT MAY BENEFIT THEM,
BY BEGETTING A FONDNESS FOR BOOKS—THE ANTIDOTES
OF IGNORANCE, OF IDLENESS, AND VICE[42]

Mayne Reid's novels sometimes read like a natural history lesson. If Eliot's was the eye which was to pick out the 'ailanthus trees' and 'flaming cardinal birds' of his childhood, and which continued to interest itself in flora from *Eliot's Floral Magazine*, and the 'moonflower' of an early 'Song' through the more famous lilacs and hyacinths of *The Waste Land* to the 'sunflower' and 'clematis' of 'Burnt Norton' with its rose garden, then that eye received training in the pages of Mayne Reid, pages overgrown with botanical names.[43] As such a flowery writer, Mayne Reid is a curiosity; but his style and subject-matter went considerably beyond what is demonstrated by this passage and provided for T. S. Eliot a stimulus for his interest in primitive society, consecrated for literary use some of the most important landscapes of his poetry, and contributed towards the formation of the poet's modes of thought. *The Boy Hunters* in particular provided a beginning for material which would form part of *The Waste Land*. Such a 'beginning' is of course a fiction. If his reading was crucially important to Eliot as a poet, then 'interpenetration and metamorphosis' equally crucially rearranged what he read. But unless poems are genetically programmed into the future poet, the reading of Mayne Reid may serve as an approximate beginning for *The Waste Land*. It is doubtful if we can go any further back.

For nine months of the year Eliot's fastest escape route was literature. Mayne Reid, like Kipling and Stevenson, could transport him from Locust St. to that Wild West just beyond St Louis and to exotic primitive landscapes far from all metropolitan life. Wildness was one of Mayne Reid's specialities as shown by such lengthy titles as *The Bush Boys or the History and Adventures of a Cape Farmer and his Family in the Wild Karoos of Southern Africa* (1856) or *The*

---

[42] *The Boy Hunters*, Dedication.
[43] See preface to E. A. Mowrer, *This American World* (London: Faber & Faber, 1928), p. xiv.

*Forest Exiles or the Perils of a Peruvian Family amid the Wilds
of the Amazon* (1855) or *The Desert Home or the Adventures of
A Lost Family in the Wilderness* (1852). And the sort of adventures
which happened in such surroundings were all that one would expect.
Cannibal fish and anacondas are second nature to the 'Forest Exiles',
one of whose most exciting adventures involves a boy being chased
by jaguars in 'the wilds of the Amazon' where eventually the boy
escapes.[44] Eliot is in the same territory in 'Up the Amazon', and it
is a territory to which, years later, he would return in his poetry
through the importation of Amazonian jungle into London maisonette,
when, in 'Whispers of Immortality', 'The couched Brazilian jaguar /
Compels the scampering marmoset.' The Brazilian landscape which
Mayne Reid presented as remote and violent, and which Eliot would
later read about as the homeland of 'some of the lowest living
men'[45] is present also in the violent world of 'Sweeney Among the
Nightingales' when 'The circles of the stormy moon / Slide westward
toward the River Plate'. The angrily potent world of the primitive
land which confronts the civilized metropolitan, a confrontation
which fascinated Eliot whether in *Tarzan of the Apes* or *Heart of
Darkness* or the person of Celia in *The Cocktail Party* or Gomez
in *The Elder Statesman*, is a world seen in Mayne Reid's novels where
similar tensions exist.[46] Such an opposition between discontented
civilization and primitive life can be seen as one of the essential
elements of the American pioneer situation. Certainly it is a constant
factor throughout Mayne Reid's work. Often, as in *The Maroon*,
a Jamaican tale, there are more details about the primitive past, about
cannibalism, and about primitive rituals. The reader learns about
'dancing the Congo dance, to the stirring sounds of the *goombay*
and *merriwang*'.[47] Details of human sacrifice and cult practices are
used to heighten his excitement. The 'horrid art' of the 'Obeah-man'
who specialized in the resurrection of the dead performs this function
in *The Maroon*.

[44] Mayne Reid, *The Forest Exiles* (London: David Bogue, 1855), pp. 287 and
371.
[45] E. B. Tylor, *Primitive Culture*, 3rd edn. (London: John Murray, 1891; 2
vols.), i. 242. Eliot cites Tylor's book in IPR (1913), using the third edition, as shown
by Piers Gray, *T. S. Eliot's Intellectual and Poetic Development 1909–1922* (Brighton:
Harvester, 1982), p. 136 n. 34.
[46] See 'Contemporanea', *Egoist*, June–July 1918, p. 84.
[47] Mayne Reid, *The Maroon* (London: Hurst and Blackett, 1862; 3 vols.)
i. 5–6.

I cannot here enter into an explanation of the mysteries of Obi, which are simple enough *when understood*. I have met it in every land where it has been my lot to travel; and although it holds a more conspicuous place in the social life of the savage, it is also found in the bye-lanes of civilization.

The reader, who may have been mystified about its meaning, will perhaps understand what it is, when I tell him that the *obeah-man* of the West Indies is simply the counterpart of the 'medicine man' of the North-American Indians, the 'piuche' of the South, the 'rain-maker' of the Cape, the 'fetish man' of the Guinea coast, and known by as many other titles as there are tribes of uncivilized men.

It is the *first dawning of religion on the soul of the savage* . . .[48]

In the Eliot who read Mayne Reid the seeds were sown which would come to fruition in the Eliot who would read widely in the history of religion, praising, for example, Durkheim's anthropological classic, *The Elementary Forms of the Religious Life*, as 'perhaps one of the most significant, and . . . one of the most fascinating, of books on the subject of religion which have been published during the present century'.[49] But a much more direct outcome of Eliot's youthful literary taste is seen in an early short story.

This island, he found out afterwards, was Matahiva, in the Paumota group. Not long after the captain was there the French got hold of it and built a post there. they [*sic*] educated the natives to wear clothes on Sunday and go to church, so that now they are quite civilized and uninteresting.

As the first white man ever to visit Matahiva, Captain Jimmy Magruder is received with much honour. He arrives there after a shipwreck and swims ashore. The captain then falls asleep in a thicket of tari-bushes, but wakes to find himself being carried on a litter by the islanders as part of a procession including priests bearing bowls of foul-smelling incense and 'a little mob of men beating bhghons (a sort of cross between tin pan and gong) and chanting monotonously'.[50] In its location, Eliot's 'The Man Who Was King' looks towards Mayne Reid's *The Flag of Distress: A Tale of The South Sea*. The story is also located in Stevenson country. Eliot locates it in Polynesia, and, more particularly, the use of the name 'Paumota' points towards the Taumotu Islands, called by

[48] Mayne Reid, *The Maroon*, i. 18–19.
[49] 'Durkheim', *Saturday Westminster Gazette*, 19 Aug. 1916, p. 14.
[50] 'The Man Who Was King', *Smith Academy Record*, June 1905, pp. 1–2.

Stevenson 'The Paumotus' in his *In the South Seas*, and mentioned in *The Ebb-Tide*, published at the height of Stevenson's popularity, when Eliot was six. Nodding towards both Kipling and Stevenson, Eliot's story seems to derive its main interest in primitive ritual from Mayne Reid. Into this interest, though, Eliot has injected a humour scarce in the pages of the Victorian writer. The natives are interesting, but also funny, bhghons and all. Later, Eliot would study primitive rites more seriously, but it is important to remember that he could also see them as ridiculous or foolish, and make mischief with them. His introduction to savage life came from popular culture, not from anthropology, and his attitudes towards the savage could shift from fascination to amusement. In this, he was typical of his time. The editor of the *St. Louis Globe-Democrat* in Eliot's youth, for instance, was clearly fascinated by primitive man, printing articles such as a long piece on the Dyak head-hunters to whom Eliot would several times refer in his later writing.[51] But always, as well as genuine interest, there goes a tone of amusement. 'Dancing "as she is taught" by French masters is little in vogue with the natives of New South Wales', begins a piece on the aboriginal skeleton dance, describing how Australian natives paint skeletons on their bodies and dance round bright fires.[52] But such an interest could lead to a deeper concern. Eliot's childhood story of a sea captain shipwrecked on Matahiva, adopted by natives as their divinely bestowed king, but then threatened with death because of an inability to perform magic feats, paved the way for his interest in more serious killings of the king.

'I am still an American in some respects and an Englishman in others,' Eliot told the *New York Times* in 1958, and repeated the next year that his poetry was 'a combination of things. But in its sources, in its emotional springs, it comes from America.'[53] Certainly as he became more ostensibly English, Eliot liked it to be remembered that he was American also. In 1928, just after joining the Church of England and taking British nationality, he made his famous pronouncement about being 'classicist in literature, royalist

[51] 'The Wild Men of Borneo' (unsigned article), *St. Louis Globe-Democrat*, Sunday Morning, 31 Jan. 1897, p. 24.
[52] 'The Skeleton Dance' (unsigned article), ibid. 7 Feb. 1897, p. 38.
[53] T. S. Eliot, 70 Today, Concedes He Looks on Life More Genially' (unsigned article), *New York Times*, 26 Sept. 1958, p. 29; 'The Art of Poetry', *Paris Review*, Spring/Summer 1959, p. 70.

in politics, and anglo-catholic in religion'.[54] But in the same year he prefaced E. A. Mowrer's *This American World*, reminding his readers that 'I am myself a descendant of pioneers.'[55]

Various factors of Eliot's life must have encouraged him to see his own background in terms of the world of Mayne Reid's novels. When he wrote, again in 1928, that his grandmother had 'shot her own wild turkeys', this was an action straight out of the pages of his boyhood favourite, an action mentioned several times in Mayne Reid's books, but perhaps most noticeably in *The Boy Hunters* where one of the chapters is entitled 'A Wild-Turkey Hunt', an event thought worthy of illustration. Eliot was later forced to admit that 'grandma had never shot a wild turkey', adding, 'I'm sorry she didn't.'[56] Without doubt, *The Boy Hunters* is the one particular Mayne Reid novel which we can be sure that he read. The Mississippi, the 'Big River', is a part of this book, whose first words are 'Go with me to the great river Mississippi', and whose quest begins with a journey to St Louis, and later involves a confrontation with Red Indians.[57] The nature of that quest is made plain in the book's original title, *The Boy Hunters or Adventures in Search of A White Buffalo*. On the hunt out of St Louis the boys encounter various perils, including alligators, Indians, and the dangers of the desert.

The stories of Uncle Remus gave Pound the nickname for Eliot, 'Old Possum', but Mayne Reid's books were also very popular in late nineteenth-century America and beyond. Possums, as curious American animals, are common in the pages of these novels. When the boy hunters return to a possum, after chasing a lynx with their dog, they are in for a surprise.

To their astonishment no 'possum was there—neither in the tree, nor the briar-patch beside it, nor anywhere! The sly creature had been 'playing 'possum' throughout all that terrible worrying . . .[58]

Possums perform such Macavity-like tricks elsewhere in the pages of Mayne Reid's novels.[59] The title of this chapter of *The Boy Hunters* is 'A Cunning Cat and a Sly Old 'Possum'.

[54] *FLA*, p. 7. [55] 'Preface' to Mowrer (B8), p. xiii.
[56] Ibid., p. xiv; *The Boy Hunters*, plate facing p. 165; 'Address', *From Mary to You*, Dec. 1959, p. 135.
[57] *The Boy Hunters*, pp. 1, 20, 84–94. [58] Ibid., p. 214.
[59] See, e.g., 'The Old " 'Possum" and her Kittens', ch. XXXVIII of Mayne Reid's *The Desert Home* (London: David Bogue, 1852), esp. p. 405.

Various features of this book seem to have impressed Eliot either in themselves or else because, as so often happened in his reading, they provided a pre-existing expression for his own experience and sensations, but one which he could adapt for his special purposes. The general tone of Mayne Reid's more solemn side is reflected in one of the most memorable sections of this book, whose subject forms the frontispiece in the original 1853 edition. This is from chapter eleven, 'The Chain of Destruction', where the boy hunters see a cycle of death in which a humming-bird (singled out as a rare sight) is eaten by a tarantula which is eaten in turn by a lizard. This 'chain of destruction' continues, and ends with a kite which is attacked by an eagle. Finally one of the boys shoots the eagle with his rifle. 'This was the last link in the chain of destruction.'[60]

Such a grim view of life as he encountered in his youth had its outcome in Eliot's own thought. One of his earliest surviving drawings is of a hawk swooping down on a rabbit.[61] Eliot's first published story, 'The Birds of Prey', dating from January 1905, tells of a man almost eaten by vultures.[62] The universality of death is seen in the cyclic form of some well-known lines in 'Marina'. With its particular reference to the American landscape in the original manuscript, and its themes of adventure at sea and half-seen faces, this poem looks back to Eliot's childhood and youth as well as looking to other literature which cloaks these experiences, so that his own actual voyages off the New England coast have been given literary sanctification.[63] But one of the alterations which Eliot made in working towards the final version of his own catalogue of destruction was to change a line from referring to a peacock to reading as it now does, 'Those who glitter with the glory of the hummingbird, meaning / Death'. Given the way in which Eliot's mind worked it is likely that this change was made because this poem about a lost child was filled with echoes and hints of his own American childhood, and because the form of this particular passage reminded him of something which lay behind it, far in the past, a passage which his own mind had been 'magnetized' to pick up: ' . . . Listen to his whirring wings, like the hum of a great bee. It is from that he takes his name of "humming-bird." See his throat,

---

[60] *The Boy Hunters*, p. 140.   [61] With *Fireside* material (Houghton).
[62] 'The Birds of Prey', *Smith Academy Record*, Jan. 1905, pp. 1–2.
[63] 'Marina' MS (Bodleian).

how it glitters—just like a ruby!'[64] The glittering of this humming-
bird, though, is far from the only aspect of wild life in the book which
was incorporated into Eliot's own literary world.

'Coriolan' is another poem much preoccupied with childhood,
though here the child is apparently a man. In Shakespeare's play we
remember how enraged Coriolanus became at Aufidius's taunt of
'Boy!'[65] Eliot's Coriolanus, for all the cataloguing of armaments in
mock- or not so mock-Homeric style, is very much a boy with his
crying of 'Mother mother'. Adolescent and childish sexuality clearly
interested the Eliot who wrote 'Dans le Restaurant', who possessed
in his own library a work on the subject, who had read in typescript
the 1936 work *Friendship-Love in Adolescence* by his friend,
Nicholas Mikhailovich Iovetz-Tereshchenko, and who discussed such
matters in relation to the childhood of his own most admired poet,
Dante, in what he described to Pound as an autobiographical
context.[66] But if 'Coriolan' relates to this interest of Eliot's, it also
relates to the reading of his own childhood, and to such extended
passages in *The Boy Hunters* as that where, in the nocturnal forest
stillness, the boys hear tree frogs and cicadas cry, and watch the
fireflies which presage a rainstorm.[67] However much overlaid with
Virgilian and other references, the basis of one of the landscapes and
atmospheres in 'Coriolan' is essentially that of Mayne Reid and the
southern states of America. Here again we see the Eliot who acquired
Mayne Reid's liking for flora and fauna, as we meet Eliot's frogs,
his 'fireflies' which 'flare against the faint sheet lightning', and his
birds and flowers among the still trees where

The small creatures chirp thinly through the dust, through the night.
O mother
What shall I cry?

Here the fireflies are part of a wild landscape associated with the
protective simplicity of childhood.

[64] *The Boy Hunters*, p. 110.
[65] *Coriolanus*, v. vii. 104, in Shakespeare, *Complete Works*, ed. P. Alexander
(London: Collins, 1951), p. 868.
[66] *SE*, p. 273; see Robert Crawford, 'Eliot's Generosity: Lindsay's Humanity',
Balliol College Annual Record, 1983, pp. 43–7; letter to Pound, 9 Dec. 1929
(Beinecke).
[67] *The Boy Hunters*, pp. 241–2.

It would be foolish to wonder whether or not Eliot was aware of all such connections. Certainly he was aware of the way in which apparently unlikely material entered his poetry.

And I should say that the mind of any poet would be magnetized in its own way, to select automatically, in his reading (from picture papers and cheap novels, indeed, as well as serious books, and least likely from works of an abstract nature, though even these are aliment for some poetic minds) the material — an image, a phrase, a word — which may be of use to him later. And this selection probably runs through the whole of his sensitive life.[68]

One of the commonest landscapes in Mayne Reid's novels is the desert. 'There is a great desert in the interior of North America. It is almost as large as the famous Saära of Africa . . . Fancy a desert twenty-five times as big as all England!' The reader encounters varieties of landscape within the desert, ranging from oases to 'large tracts of mountainous and hilly country' to great plains of white sand and 'other plains, equally large, where no sand appears, but brown barren earth utterly destitute of vegetation'. What vegetation there is includes most commonly the artemisia of the *'sage prairies'* which is 'a stunted shrub with leaves of a pale silvery colour. In some places it grows so thickly, interlocking its twisted and knotty branches, that a horseman can hardly ride through among them.'[69] Such a landscape would haunt Eliot.

> What are the roots that clutch, what branches grow
> Out of this stony rubbish?

But (no doubt with biblical reinforcement) it was the story of the search for water in the desert which made an even deeper impression. An extended sequence in *The Boy Hunters* describes the crossing of the desert, contrasting the attitude of the boys with that of seasoned travellers. The boy hunters set off confident of reaching water within a day, but are reproached by the author for not knowing that 'the fear of thirst is . . . the greatest of all terrors'.

Our young hunters felt but little of this fear. It is true they had, all of them, heard or read of the sufferings that prairie travellers sometimes endure for want of water. But people who live snugly at home, surrounded by springs, and wells, and streams with cisterns, and reservoirs, and pipes, and hydrants, and jets, and fountains, playing at all times around them, are prone to

---

[68] *Use*, p. 78.   [69] *The Desert Home*, pp. 1–3.

underrate these sufferings; in fact, too prone, might I not say, to discredit everything that does not come under the sphere of their own observation.[70]

The perception of something awful beyond normal observation is what would attract Eliot to some of Kipling's stories and to Conrad's *Heart of Darkness* and is what would become one of the major themes of his own poetry. In *Heart of Darkness*, Marlow is one who has deviated from the routine life of a city clerk into the life of the African jungle, but Eliot, though he can set the jungle of India against his City, tends to use not so much the jungle but the desert as the counter to the city in *The Waste Land*. This may be partly the result of the romantic tradition, of 'Ozymandias' or of 'The City of Dreadful Night', but it is also because to wander off from one's normal course into the terror of the desert was one of the images which was most familiar to Eliot, an experience at once open to all men, from biblical times to the twentieth century, and one which he could adapt for his own purposes. 'Even', wrote Mayne Reid, 'the rude savage and the matter-of-fact trapper often diverge from their course, impelled by a similar curiosity.'[71] The terror of the desert is the terror hidden from the dwellers in cities during their normal routine, a terror which comes out of being uprooted from the familiar.

I was saying, then, that people who live at home do not know *what thirst is*: for *home* is a place where there is always water. They cannot comprehend what it is to be in the desert without this necessary element. Ha! *I* know it; and I give you my word for it, it is a fearful thing. Our young hunters had but a fair idea of its terrors.[72]

Mayne Reid's chapter 'A Night in the Desert', which follows in *The Boy Hunters*, describes a journey across the desert and the landscape which the hunters pass. This includes, for instance, the *baranca*.

It was precipitous on both sides, with dark jutting rocks, which in some places overhung its bed. There was no water in it to gladden their eyes; but even had there been such, they could not have reached it. Its bottom was dry, and covered with loose boulders of rock that had fallen from above. . . . They chose the path that appeared to lead upward . . . but still the fissure, with its steep cliffs, yawned below them, and no crossing could be found. . . . with feelings almost of despair . . . They did not sleep even for a moment. The agonising pangs of thirst as well as the uncertainty of what was before them on the morrow kept them awake . . . a deep buffalo-road

[70] *The Boy Hunters*, pp. 272–3.
[71] Ibid., p. 275.   [72] Ibid., p. 274.

. . . was no termination to their sufferings, which had now grown more acute than ever. The atmosphere felt like an oven; and the light dust . . . enveloped them in a choking cloud . . . It was of no use halting again. To halt was certain death — and they struggled on with fast-waning strength, scarcely able to speak to one another. Thirst had almost deprived them of the power of speech! . . . Their eyes were thrown forward in eager glances — glances in which hope and despair were strangely blended.

The grey, rocky bluff that fronted them, looked parched and forbidding. It seemed to frown inhospitably upon them as they drew near.

'O brothers! should there be no water!'

For the Boy Hunters relief comes in the form of 'a prairie spring' found 'round a point of rocks'.[73] But for Eliot in *The Waste Land* such a goal is never reached, though it remains longed for in that search so strongly reminiscent of the search in his boyhood reading, though now endowed with an adult significance and a despair which would have met with the approval of the James Thomson who had written of William Blake in the 'desert of London town'.

> Here is no water but only rock
> Rock and no water and the sandy road
> The road winding above among the mountains
> Which are mountains of rock without water
> If there were water we should stop and drink
> Amongst the rock one cannot stop or think
> Sweat is dry and feet are in the sand
> If there were only water amongst the rock . . .
>   If there were water
> And no rock
> If there were rock
> And also water
> And water
> A spring
> A pool among the rock
> If there were the sound of water only . . .

It would be ridiculous to claim that these lines could have been written by a twelve-year-old boy. But underneath the personal agony sieved through a reading of Durkheim, Lévy-Bruhl, Frazer, Spencer, and Gillen, and the host of other anthropological writers on deserts whom Eliot read, there lies, along with repeated readings of the Bible,

[73] Ibid., pp. 308–12.

this simple quest pattern in Mayne Reid whose novels so impressed
Eliot as a child that he imitated them in some of the articles written
in the *Fireside* when he was eleven.

Supporting such a contention is the fact that the other great underpinning quest of *The Waste Land* was also well known to Eliot as a
child. At Smith Academy, St Louis, he was made to read some of
the *Idylls of the King*, but more important is the fact that at the age
of eleven or twelve, that is at around the time when he was reading
Mayne Reid's novels, he was also reading what was even then his
favourite book: a children's edition of Malory.[74] Malory's great
Arthurian tapestry, shored together from so many fragments, remained
one of the books dearest to the author of *The Waste Land*. In 1934
Eliot wished that there were available three versions of the book.
First, a children's edition, such as that which he had read at eleven
or twelve. Secondly, a cheap edition of the text, and thirdly, and
this is most important, an edition of the work done in a scholarly
fashion and with an extensive commentary by someone such as Jessie
Weston or Jane Harrison.[75] Here again we see the stratification which
is so important to Eliot's development. 'Not only the title [he wrote
of *The Waste Land*], but the plan and a good deal of the incidental
symbolism of the poem were suggested by Miss Jessie L. Weston's
book on the Grail Legend.'[76] But beneath his reading in Weston,
Harrison, Frazer, and other anthropologically influenced scholars
there lay once again his childhood reading, this time in Malory.

To think of *The Waste Land* as embodying elements from Eliot's
own childhood may seem strange; but those cries,

> Twit twit twit
> Jug jug jug jug jug jug
> Drip drop drip drop drop drop drop
> Co co rico co co rico

look not only towards past literature and towards the sympathetic
magic of the ecstatic cries in primitive ritual; they also come from
the Eliot who was, as he told the American Academy, a devoted
birdwatcher as a child.[77] The Eliot who provides his readers with

---

[74] 'On Teaching the Appreciation of Poetry', *Critic* (Chicago), Apr.–May 1960,
p. 13.
[75] 'Le Morte Darthur', *Spectator*, 23 Feb. 1934, p. 278.
[76] CPP, p. 76.   [77] 'The Influence of Landscape' (C639), p. 422.

the Latin name of the hermit thrush in a note to line 357 of *The Waste Land* when he quotes from Chapman's *Handbook of Birds of Eastern North America*, a book which he received as a present when aged fourteen, is the same Eliot who enthused in 1932 correspondence about New England birdlife when his return to Cape Ann brought bursting forth in Eliot one of his boyhood enthusiasms, as shown in the nervously excited 'Cape Ann' poem, celebrating that birdlife left behind.[78] In 'Burnt Norton' a similar quickening is initiated when 'the bird' is introduced. Significantly, it is this thrush which leads Eliot towards a world of children and a mysterious door, thought by many to be related to a door in the wall of Eliot's own childhood garden in St Louis.[79]

It is hard to overestimate the importance of Eliot's childhood reading and the experience of which it formed part. From the first, he was impressed by fragments of quotation, such as four lines from 'The Vanity of Human Wishes' which in 1930 he recalled reading thirty years earlier as the epigraph to *Ivanhoe*'s last chapter.[80] The lines from *Julius Caesar* (which Eliot read and disliked at school),

> Between the acting of a dreadful thing,
> And the first motion, all the interim is
> Like a phantasma or a hideous dream

often linked by critics to the falling of the Shadow in 'The Hollow Men' appear in an epigraph from another book which Eliot read when he was eleven, as does the phrase 'O dark, dark, dark!'[81] The name Prufrock comes from St Louis.[82] So, more importantly, does that of the figure whom Eliot would later use as a sort of city savage: Sweeney.

---

[78] *CPP*, p. 79; John J. Soldo, *The Tempering of T. S. Eliot* (Ann Arbor, Michigan: UMI Research Press, 1983), p. 6; letter to Dobrée, SS. Simon & Jude, 1932 (Brotherton); *CPP*, p. 142.

[79] Eliot mentions the door in 'Address' (C637), p. 134; see also Ronald Bush, *T. S. Eliot, A Study in Character and Style* (New York: Oxford University Press, 1984), pp. 132–3.

[80] 'Introductory Essay' to Samuel Johnson, *London: A Poem* (London: Frederick Etchells and Hugh Macdonald, 1930), p. 9.

[81] See 'Sherlock Holmes' (C283), pp. 552–6; Anna Catherine (*sc.* Katherine) Green, *The Leavenworth Case* (London: Strachan, 1884), pp. 73 and 445; see also Grover Smith, *T. S. Eliot's Poetry and Plays: A Study in Sources and Meaning*, 2nd edn. (Chicago: University of Chicago Press, 1974), p. 102; and Eliot's 'On Teaching' (C640), p. 14.

[82] Stephen Stepanchev, 'The Origin of J. Alfred Prufrock', *Modern Language Notes*, June 1951, pp. 400–1.

MAKE NO MISTAKE.

Doctor F. L. Sweany.

Fig. 2. *Doctor Sweany*

Doctor F. L. Sweany, of northwest corner, Broadway and Market Street, St Louis, advertised daily in the *St. Louis Globe-Democrat*.[83] His speciality, the standard form of the advertisement declared, was 'Nervous Debility'. Here is the doctor Prufrock required, as his advertisement shows. 'Is Your Body and Brain Fatigued? ARE YOU LACKING IN ENERGY, STRENGTH AND VIGOR . . . MEN WHO ARE WASTING AWAY! Do you want to be cured?' Eliot was so impressed as a child with this advertisement that he drew his own copy in the *Fireside*, reproducing the doctor's name, face, and rubric with its warning to seek consultation 'when others fail'.[84] Eliot even reproduced the hirsute Dr Sweany's enormous beard. Later, 'Apeneck Sweeney' would be involved in Eliot's examination of what 'manliness' might mean, as would the failure, Prufrock, worrying that 'They will say: "But how his arms and legs are thin!".' If Dr Sweany merged in Eliot's later years with a Steve O'Donnell (a Boston pugilist with another Irish surname) and a tough bartender, the essential concern with manliness was only heightened further.[85] Lack of 'energy,

---

[83] In the first six months of 1897, e.g. Sweany's advert appeared in the *Globe-Democrat* virtually every day. The rubric, 'When Others Fail Consult Doctor F. L. Sweany' always surrounded Sweany's head and shoulders.

[84] *Fireside*, no. 4, 2 Jan. 1899 (Houghton).

[85] See, e.g., Conrad Aiken, 'King Bolo and Others', in Tambimuttu and Richard March, eds., *T. S. Eliot: A Symposium* (New York: Tambimuttu and Mass, 1948), p. 21; see also *Facsimile*, p. 125 n. 6.

strength, and vigor' becomes chronic in 'Gerontion' with its 'sleepy corner'. But there is already too much sleep in the etherized 'Prufrock', getting nowhere slowly in his run-down city streets. Eliot's Dr Sweany cured insomnia. Though the Sweeney of the poems is sometimes a caricature of the 'macho', Eliot has some sympathy for him. In *Sweeney Agonistes* he cannot sleep for nightmares.

It is not hard to see in 'Prufrock' a development of that drab urban landscape present to him in what he remembered as 'the large industrial city of St Louis', to which, by 1910, Eliot had added considerable knowledge of Boston.[86] But it is important, also, to remember the elegant urbanity of the interior of 'the room' beyond the 'sprinkled streets'. Here again Bostonian society may have played its part, and Eliot grew up in St Louis in a family with four considerably older sisters.[87] Sisters, however, are seldom objects of erotic fear. Yet Eliot was prepared in St Louis for the sort of world which Prufrock would perceive. The first detective story which Eliot read, aged about eleven, was A. K. Green's *The Leavenworth Case*, a murder story centring around two New York beauties.[88] Thirty years later, cheerfully criticizing it for being 'simply popping over with sentiment', Eliot singled out a passage to exemplify what he meant. Green's words to which Eliot added an ironic 'etc.', read:

'Seated in an easy chair of embroidered satin, but rousing from her half-recumbent position . . . I beheld a glorious woman. Fair, pale, proud, delicate; looking like a lily in the thick creamy-tinted wrapper that alternately clung to and swayed from her richly moulded figure; with her Grecian front, crowned with the palest of pale tresses lifted and flashing with power . . . I held my breath in surprise, actually for the moment doubting if it were a living woman I beheld . . . etc.'

Such a fearful half-recumbent goddess foreshadows the female inhabitants of Prufrock's world where pillows are settled and shawls thrown off. In 1929 Eliot breaks the spell of *The Leavenworth Case* completely by pointing out that 'this slumberous blonde is merely Mary', but this is exactly what the awestruck Prufrock cannot do.[89] His women are nameless and awe-inspiring. To name in Eliot's early poems is normally to make ridiculous (J. Alfred Prufrock, Aunt Helen Slingsby, Professor Channing-Cheetah). As in the later work, the unnamed or unnameable is most fearful and fascinating. Mary

---

[86] 'President Wilson' (C40a), p. 140.   [87] Ackroyd, p. 16.
[88] 'Sherlock Holmes' (C283), p. 552.   [89] Ibid., p. 555.

Leavenworth, in the passage from which Eliot quotes, is described as a 'pythoness'.[90] In her rich New York apartment she is Eliot's first Cleopatra. Her sister, faintly lit by a glimmering gas-jet, sits in a marble apartment.[91] However, as Eliot's reminiscence points out, these women are harmless. Part of the comedy of Prufrock's situation comes from his crucial tendency to envisage women as remote and menacing, a foible he shares with the junior law partner who narrates *The Leavenworth Case*. This book forms another part of Eliot's clerkly world.

The urban landscape of the early poems is not exclusively sordid. Seediness is played off against the rich female-ruled townhouse interiors of 'Prufrock' and 'Portrait of a Lady' whose 'tastefulness' is sent up. No doubt these were based partly on the Bostonian home of Adeline Moffat, and of the ageing Isabella Stewart Gardner whose home he visited for art classes in 1910 (she had been to Paris in her youth).[92] A museum-house, its rooms whimsically composed of artworks arranged like bric-à-brac, with carved ceilings and elaborate candelabra, survives intact to this day. But Eliot had been 'magnetized' to pay attention to such female-dominated urban interiors both by his reading about the sentimentalized world of a fashionable metropolis and by his own childhood with his four sisters among the 'sprinkled streets' of St Louis. Prufrock carries his streets with him. They 'follow' like his own 'tedious argument', rather than *leading*, as normal streets do. St Louis, as Eliot acknowledged, remained insistently a part of his own life.[93] It is often assumed that Prufrock was the fate which lay in wait for Eliot, but the young Eliot in St Louis, writing in the *Fireside* his regular mischievous columns of 'Gossip' (the thing Prufrock fears so much) also lay in wait for Prufrock.

Even when Eliot seemed furthest from his mid-western Unitarian upbringing as, for instance, when he attended Masaharu Anesaki's lectures on Japanese Buddhism in 1913–14 at Harvard, went to the

---

[90] *The Leavenworth Case*, pp. 59–60.   [91] Ibid., p. 123.

[92] Matthews, p. 26, mentions Moffat and Eliot's paying his respects to 'Mrs. Jack Gardner, the Boston monument'. Eliot's 1910 notes on his Harvard course, Fine arts 20b (Houghton), include notes on paintings in the collection of a Mrs Gardner's in whose house the class occasionally met. Eliot wrote to Isabella Stewart Gardner from Oxford and London in 1915. See also Anon., *Guide to the Collection, Isabella Stewart Gardner Museum* (Boston: The Trustees of the Gardner Museum, rev. edn., 1980), pp. 6–11.

[93] 'The Influence of Landscape' (C639), p. 422.

Buddhist Society in Oxford in 1915, and around 1922 came close
to becoming a Buddhist himself, he was only taking further an interest
of his childhood days when he discovered Sir Edwin Arnold's *The
Light of Asia*.[94] 'It is a long epic poem on the life of Gautama
Buddha: I must have had a latent sympathy for the subject-matter,
for I read it through with gusto, and more than once.'[95] Such a
'latent sympathy' for eastern subject-matter was no doubt encouraged
by another childhood enthusiasm, this time for Rudyard Kipling
whose *Barrack Room Ballads* and *Plain Tales from the Hills* Eliot
remembered reading as a boy.[96] Kipling, too, connected the city
not with the savage, but with an often violent world of jungles
and remote mountains. *Kim*, a book of which Eliot much later
pronounced himself 'very fond', appeared in 1901.[97] Its themes
of a quest for salvation, talk of gathering ethnological data, and
particularly its Buddhist themes of the wheel of creation, escape from
the body and from 'The illusion of Time and Space and of Things'
would be of great importance to the older Eliot.[98]

It has become customary in writing of Eliot's childhood to emphasize
particular aspects which in retrospect might point to his conversion
to Christianity, such as his being smuggled into Mass by Annie Dunn,
his Catholic nurse or the importance of his mother's religious poetry
and concerns.[99] With a poet who attached such great importance
to his family, this is justified. Yet it is equally important to see that
when Eliot rebelled against the society and religion of his upbringing
not only in moving from Unitarianism towards Buddhism, but also
in setting the rituals of primitive man against the Mass, the tom-
tom against Chopin, and filthy scraps in city streets against the
conventionally poetic landscape, he was also developing childhood
interests, frequently combining the savage and the city as they had
been linked in the St Louis of his boyhood.[100]

---

[94] Eliot's full notes on Anesaki's course are in Houghton. He mentions visiting
Oxford's Buddhist Society in his letter of 4 Apr. [1915] to I. S. Gardner; he stated
that around 1922 he inclined to Buddhism — see Stephen Spender, *Eliot* (Glasgow:
Collins, 1975), p. 26.

[95] *OPP*, p. 42.

[96] 'The Unfading Genius of Rudyard Kipling', *Kipling Journal*, Mar. 1959, p. 9.

[97] 'T. S. Eliot, 70 Today . . . ', *New York Times*, 26 Sept. 1958, p. 29.

[98] Rudyard Kipling, *Kim* (1901; rpt. London: Macmillan, 1956), p. 411.

[99] See, e.g., Lyndall Gordon, *Eliot's Early Years* (Oxford: Oxford University
Press, 1977), pp. 3–6.

[100] See 'The Ballet', *Criterion*, Apr. 1925, p. 441; *CPP*, pp. 18–19 and 22.

It would be wrong to leave that childhood, however, without looking at one element linking the savage to the city: the sea. Captain Jimmy Magruder in 'The Man Who Was King' lands on Matahiva by being shipwrecked. Another early story, 'A Tale of a Whale', describes a voyage chasing a whale through the South Seas from Tanzatatapoo Island to Honolulu.[101] The Eliot family summered annually from 1893 at Gloucester, Massachusetts.[102] New England was Eliot country. The poet's ancestors had played their part in the hanging of the Salem witches, and as a student Eliot recalled his fascination with ship illustrations and with artefacts brought back from as far as China and the Levant.[103] The treasures of maritime Salem spoke of the sea, but also of the places beyond it where the young Eliot could find African pirates and sea serpents.[104] The sea could lead easily into Mayne Reid territory. One of the most famous and striking features about the Peabody Museum in Salem was its ethnological collections, containing artefacts from the Pacific, Africa, America, India, and the Far East.[105]

Eliot summered with the family in the little village of East Gloucester, about a mile round the bay from Gloucester itself, a venue which was just beginning to become a fashionable resort.[106] But it was the business and tradition of the fishing port that impressed the boy who looked out day after day on the wild Atlantic, on the low drifting fog, hearing the rhythm of the waves. Here it was that he peered, aged ten, into a rockpool, as well as classifying algae on the shore. Both events would be recalled in 'The Dry Salvages'.[107] The great events of the fishing port impressed him. All his life he remembered being at the launch of the *Rob Roy*, a Gloucester schooner which first put out to sea in 1900.[108] In East Gloucester Eliot was familiar with the great Gorton-Pew fisheries, with their unloading, filleting,

---

[101] 'A Tale of a Whale', *Smith Academy Record*, Apr. 1905, pp. 1–3.

[102] Ackroyd, p. 22.

[103] 'We didn't burn them, we hanged them,' Eliot's note in his edition of *Literary Essays of Ezra Pound* (London: Faber & Faber, 1960), p. 391; 'Gentlemen and Seamen', *Harvard Advocate*, 25 May 1909, p. 115.

[104] Ibid., pp. 115–16.

[105] See Walter Muir Whitehill, *The East India Marine Society and the Peabody Museum of Salem: A Sesquicentennial History* (Salem: Peabody Museum, 1949).

[106] See Joseph E. Garland, *The Gloucester Guide: A Retrospective Ramble* (Gloucester, Mass.: Gloucester 350th Anniversary Celebration, Inc.: 1973), pp. 75–91.

[107] *Use*, pp. 78–9; Eliot mentions classifying algae on the Massachusetts shores in letter to Pound of 22 Oct. [1936] (Beinecke); *CPP*, p. 184.

[108] Letter to J. J. Sweeney of 29 Sept. 1952 (Houghton).

salting, drying, boning, cutting, grinding, smoking, and boxing of
millions of pounds of fish brought back from the North Atlantic
banks by a fleet of Gloucester schooners. Photographs of the fisheries
at East Gloucester survive from around this time.[109] Such memories
of childhood help to explain why that moment of release in *The
Waste Land* comes where in the poem's City, 'fishmen lounge at
noon'.

When Eliot was a boy in Gloucester, between his sixth year in
1894 and 1905 when he was seventeen, the largest construction
project which the town had known was under way. This was the
making, out of local grey granite, of the Dog Bar Breakwater at
whose end, eventually, a bell was placed, adding its sound to the
whistling buoy southward, known as Mother Ann's Cow.[110] Mother
Ann is a rock formation at Eastern Point, at the extremity of Cape
Ann. The grey granite and the noises associated with it would possess
Eliot, who uses distinctive seamen's words such as 'rote' and 'groaner'
remembered from the New England coast of his youth, when he had
heard 'The distant rote in the granite teeth, / And the wailing warning
from the approaching headland', which he recalled in 'The Dry
Salvages'. As a child, Eliot himself took sailing lessons out on Cape
Ann, under his mother's anxious supervision.[111] He rejoiced in his
knowledge of nautical terms in early stories, and his seamanship
would stand him in good stead as a student, when he would sail
much further afield up the New England coast.[112] Such experience
is found not only in 'The Dry Salvages'. It is also concealed in *The
Waste Land* in one of those other Stevensonian phrases which Eliot
presents, possum-like, yet which is loaded with his own childhood
memories.

> At the violet hour, the evening hour that strives
> Homeward, and brings the sailor home from sea, . . .

Eliot's note relates this last line to Sappho's poem on Hesperus:
'Hesperus, bringing [back] all things which bright dawn scattered;
you bring the sheep, you bring the goat, you bring the child back

---

[109] Garland (above, n. 106), pp. 76 and 77.
[110] Ibid., p. 88.   [111] Gordon, pp. 3–4.
[112] T. S. Eliot with R. W. H. and L. M. L., tribute for 'Thomas Stearns Eliot',
Harvard College Class of 1910, Fifty-fifth Anniversary Report (Cambridge, Mass.:
Crimson Printing Company, 1965), p. 53.

to its mother.'[113] About his own lines Eliot writes, 'This may not appear as exact as Sappho's lines, but I had in mind the "longshore" or "dory" fisherman, who returns at nightfall.'[114] Hidden here is a return to Eliot's own Mother Ann and perhaps to his own mother, as well as to the Gloucester dory fishermen of his childhood, the dory being a distinctively New England vessel. When Eliot first published 'The Dry Salvages' in 1941, he included in his introductory note the statement that 'The Gloucester fishing fleet of schooners, manned by Yankees, Irish or Portuguese, had been superseded by motor trawlers.' Later this sentence, implying perhaps that the present is less heroic than the past, was dropped, but Eliot kept his suggested imaginative etymology, 'The Dry Salvages'—presumably *les trois sauvages*.[115] 'The Three Savages' because the rocks claimed many ships. But the name also hints at what could lie beyond the sea, as well as at the American past. In *The Waste Land*, the pain of returning memory, associated with Eliot's sailing days, would be expressed largely through replacing the personal with the recurring rituals of primitive men and a horrifying vision of London life. Yet one of the crucial geographies of Eliot's childhood and of his childhood imagination would also return in that poem: the familiar geography of the city, the savage, and the sea between them. 'Home is where one starts from.'

[113] Translated by Constantine A. Trypanis, ed., *The Penguin Book of Greek Verse* (Harmondsworth: Penguin Books, 1971), p. 148.

[114] *CPP*, p. 78.

[115] Headnote to 'The Dry Salvages', *New English Weekly*, 27 Feb. 1941, p. 217; *CPP*, p. 184.

II

CITY

IT was the exotic side of Eliot which first quickened to poetry. From importing into St Louis Mayne Reid's South Seas, jungles, and deserts, he moved on to the dusty, windy 'Waste' of *The Rubáiyát of Omar Khayyám* with its 'Hyacinth . . . Garden', its gloomy concerns with 'Nothing', and with life as illusion.[1] Fitzgerald's *Rubáiyát* was a cult book in late nineteenth-century America.[2] Eliot was converted, however, momentarily, to the cult, with the result that Fitzgerald's vocabulary, even when rearticulated for other ends, became his own and his personal experience was mapped on to the pattern of his reading.

Eliot devoured Fitzgerald's poem with great enthusiasm, but the *Rubáiyát* is crucial too because, after his anthropological reading, he would be able to re-utilize many of its elements in a new way. Disinfected of their vague romanticism by being plunged into the world of the scientifically observed savage, Fitzgerald's deaths and rebirths, his desert and garden would be revived using the respectably clinical (and so classical) lens of Frazer, Weston, and others. For Eliot anthropology would fill with savage rituals, and so reinstate, the romanticism of

> And those who husbanded the Golden grain,
> And those who flung it to the winds like Rain,
>   Alike to no such aureate Earth are turn'd
> As, buried once, Men want dug up again.   (XV)

Fitzgerald produced a long poem with strange notes which deal in a variety of languages with festivals celebrating the revival of spring, with comparative folklore, east and west, with the cries of birds and their literary representation.[3] In this he would have a successor.

---

[1] See Leonard Unger, *Eliot's Compound Ghost* (University Park: Pennsylvania State University Press, 1981), *passim*.
[2] Fitzgerald, *Works* (New York: Houghton Mifflin, 1887; 2 vols.), 'Dedication', i. p. v, and 'Biographical Preface', i. pp. xix–xx.
[3] Fitzgerald, *Works*, i. 76–82 (references are to the 4th edn. of the *Rubáiyát*).

Eliot became renowned for his attack on nineteenth-century
romanticism. With the 'dissociation of sensibility' theory, he seemed
to throw away the Victorians in favour of the Metaphysicals, just
as surely as he had thrown away his early 'very gloomy and atheistical
and despairing quatrains' written 'under the inspiration of Fitzgerald's
Omar Khayyam'.[4] Both castings off are illusory. It is ironic, but
probable, that the first lines of Donne which Eliot read were
contained in Fitzgerald's 'Notes' to the *Rubáiyát*. Fitzgerald is part
of the nineteenth century which Eliot buried. But he was always
digging it up.

From Fitzgerald's desert to Thomson's City of Dreadful Night was
a short journey. Thomson's poem intensified the pessimism and sense
of impotence found in Fitzgerald's. The desert reappeared in
Thomson's work as a frequent symbol for man's life. If Fitzgerald
gave Eliot a new and striking experience of poetry when he was
fourteen, Thomson's poem had a similar effect two years later, and
Eliot went on to read further in the Victorian poet's work. Again,
though, as with Mayne Reid and Fitzgerald, Thomson seems to have
been especially exciting to Eliot by being off the approved track.
Through perusing a part of his school history which he was not
required to look at, Eliot discovered when he was sixteen both *The
City of Dreadful Night* and Ernest Dowson's poems. Each he later
recalled as having struck him with the force of 'a new and vivid
experience'.[5] Eliot was partial to the idea that his admiration for
Dowson's 'Cynara' gave him the phrase 'Falls the Shadow'.[6]
Dowson's was a poetry of loss, appealing to an Eliot almost too
young to have lost anything. But Dowson would be bound up,
sometimes ironically, with later experience. His 'Chanson sans
paroles' supplied *The Waste Land*'s 'violet air', while its line, 'I wait
for a sign' anticipates 'Gerontion'. The Dowson poem's air of stillness
'laved / In a flood of sunshine' with 'no sound heard, / But afar,
the rare / Trilled voice of a bird' joins with E. B. Brownings 'The
Lost Bower' to prepare the way for 'Burnt Norton'.[7] A poem Eliot
specifically recalled, Dowson's 'Impenitentia Ultima', by mixing

    [4] 'The Art of Poetry' (C631), p. 49.   [5] 'On Teaching' (C640), p. 78.
    [6] 'Dowson's Poems' (letter), *TLS*, 10 Jan. 1935, p. 21.
    [7] Ernest Dowson, *Poetical Works*, ed. D. Flower, 3rd edn. (London: Cassell,
1967), pp. 89–90. The poems quoted are from *Verses* (1896): cf. *CPP*, pp. 37, 171–2.
For E. B. Browning, see Helen Gardner, *The Composition of 'Four Quartets'* (London:
Faber & Faber, 1978), pp. 40–1.

intense religious and sexual emotion leads towards Eliot's reading of Baudelaire. As we shall see, Dowson's presence is strongest in 'The Hollow Men' where again anthropology reinstates romanticism.[8]

But for the moment it is Eliot's introduction to Thomson's poetry which is of greater interest, since that introduction would crucially affect his view of the city. When Thomson is mentioned in connection with Eliot, it is normally as the author of *The City of Dreadful Night*, but that indefatigable source-hunter, Grover Smith, has pointed out that in the lines from 'The Hollow Men', 'Lips that would kiss / Form prayers to broken stone,' Eliot has made use of those lines from part III of Thomson's 'Art' which he was later to cite in the concluding paragraph of *The Use of Poetry and The Use of Criticism*, 'Singing is sweet; but be sure of this, / Lips only sing when they cannot kiss.'[9] The same commentator has indicated that the words at the beginning of *The Waste Land*, 'covering / Earth in forgetful snow, feeding / A little life with dried tubers', draw on a memory of two lines from Thomson's poem, 'To Our Ladies of Death', 'Our Mother feedeth thus our little life, / That we in turn may feed her with our death.'[10] Here it is plain that the suggestion of scantily renewed life leading only to the unpleasant consequence of death—or, rather, in the case of these poems, the desired consequence of death—makes the appropriation of Thomson's words appropriate for Eliot. The general setting of the Thomson poem makes it all the more suitable for Eliot's purpose since the renewal of life takes place in a poem whose speaker, not unusually for the speaker of a Thomson poem, is

> Weary of erring in this desert Life,
> Weary of hoping hopes for ever vain,
> Weary of struggling in all-sterile strife.[11]

But to the reader of Thomson another phrase in these opening lines of *The Waste Land* is also seen to be an echo of a Thomson poem with a dry, desert setting. The phrase is the famous 'memory and desire' which is to be found in stanza XIX of the second part of the narrative love poem, 'Weddah and Om-el-Bonain'. This poem

[8] Dowson's presence in 'The Hollow Men' is stressed by R. K. R. Thornton, *The Decadent Dilemma* (London: Arnold, 1983), pp. 94–6. See also ch. VI below.
[9] Grover Smith (1974), p. 101.　　[10] Ibid., p. 72.
[11] *Poetical Works of James Thomson*, ed. Bertram Dobell (London, Reeves and Turner and B. Dobell, 1895; 2 vols.), i. 235; hereafter cited as *PW*, volume, page. All selections from *The City of Dreadful Night* are from this edition (i. 122–72) and will be cited by section number only.

tells the story of two young lovers whose voices 'sprang like fountains
'mid the desert sands' (Part One, I), but whose love is blighted when
the girl, Om-el-Bonain, marries another man and her warrior lover,
Weddah, learning that he must relinquish his love, is brought very
close to death, but, recovering, receives a token from Om-el-Bonain
which makes him remember their past life in lines related not only
to parts of *The Waste Land*, but also, perhaps, to the 'warm rain'
of 'Gerontion':

> The coiled-snake Memory hissed and sprang and stung:
> Then all the fury of the storm was shed
> From the black swollen clouds that overhung;
>   The hot rain poured, the fierce gusts shook his soul,
>   Wild flashes lit waste gloom from pole to pole.   (Part Two, XI)

This return of memory coupled with the reading of a note from
Om-el-Bonain to the effect that they must part again sends Weddah
to the point of 'Peace and oblivion' (Part Two, XVI) so that he 'lay
benumbed with wounds, and would have died / Unroused' (Part
Two, XVIII). Once more it is an access of memory which, as he
wanders alone in the desert, stings him back to painful life when he

> Now demon-driven day and night pursued
> Stark weariness amidst the clamorous throng
>   Of thoughts that raged with memory and desire,
>   And parched, his bruised feet burning could not tire.   (Part Two, XIX)

But this nagging return of memory, though it will lead to a renewal
of love, will lead also to death when Weddah is buried alive by Om-
el-Bonain's jealous husband, and when she too dies grieving. Again
the context guides Eliot's use of the phrase, marrying with the themes
of painful renewal and death at the beginning of *The Waste Land*
and with the Frazerian buried corpse of Eliot's anthropological
interests; but we can see too how the later poet has removed the
phrase from Thomson's fairly frantic narrative and contained it in
a passage which is both more controlled and of more universal
application. Here as so often in his poetry Eliot is stashing away
violent emotion behind the lines of his own text, yet at the same
time moving from the personal to the impersonal, stealing as a mature
poet, rather than committing the poetaster's petty larcenies.

Eliot's use of these less well-known Thomson poems, 'Art', 'To
Our Ladies of Death', and 'Weddah and Om-el-Bonain', shows that

parts of Thomson's poetry impressed him sufficiently to lie at the back of his mind for a considerable time after he had read that poetry initially in the early years of the century. Eliot's use of these poems shows too that when he read Thomson's work he read it in considerable quantity. The most likely edition in which these three poems along with *The City of Dreadful Night* would have been available to the young Eliot was the two volume *Poetical Works of James Thomson*, edited by Thomson's friend Bertram Dobell who also contributed a lengthy memoir giving in considerable detail Thomson's biography and talking of his life as a city clerk in London.

In the late nineteenth century a central change occurred in the way that Thomson's most famous poem was read. Eliot may not have been aware of this change, but he was affected by it. It is unlikely that 'Weddah and Om-el-Bonain', 'Art' and 'To Our Ladies of Death' would have been included along with *The City of Dreadful Night* in his school anthology. More probably an enthusiasm for *The City of Dreadful Night* in his second last year at school led Eliot on to read further in Thomson. This would account for his clear memory of having read *The City of Dreadful Night* 'at sixteen', yet also associating Thomson with the time 'towards the end of my school days or in my first year or two at Harvard University'.[12] Eliot continued to read *The City of Dreadful Night* until the end of his life, long after he had himself been a London clerk.[13] When Thomson's *City* appeared it was seen as dealing with a city of fantasy, but by the time Eliot read it a change in readers' perceptions had resulted in its being regarded as a work about London, with Thomson's career seen as often paralleling that of Dante.[14]

When Eliot looked back on Thomson's work, it was with a sense of special gratitude to poets whose works had deeply impressed him during his formative years between the ages of sixteen and twenty. Among these he paid particular tribute to two Scots, the poet of *The City of Dreadful Night* and John Davidson, who wrote 'Thirty Bob a Week'.[15] Eliot remembered reading Thomson along with the

---

[12] 'On Teaching' (C640), p. 78; see Maurice Lindsay, 'John Davidson — The Man Forbid', *Saltire Review*, Summer 1957, p. 57, for Eliot's remarks in a radio tribute to Davidson.

[13] Mrs T. S. Eliot, letter to the present writer, 18 Oct. 1983.

[14] See Robert Crawford, 'James Thomson and T. S. Eliot', *Victorian Poetry*, Spring 1985, pp. 23–41.

[15] 'Preface' to John Davidson, *A Selection of his Poems*, ed. Maurice Lindsay (London: Hutchinson, 1961), p. xi.

poetry of the 1890s and, in particular, the work of John Davidson.[16] This in itself suggests that it was partly the modern urban material in Thomson which, no doubt along with the intense religious concern, Eliot found attractive. His friend Conrad Aiken admired 'the "city" celebrants, Whitman most notably, but with some attention too to such minor and different devotees as John Davidson and Henley. The city, the city—.' Another Harvard literary friend, W. G. Tinckom-Fernandez shared such an interest.[17] Thomson himself had been a Whitman devotee, and there is other evidence that educated Americans around this time thought of the poet of *The City of Dreadful Night* as a poet of London, following the biographical reading of his poem common at the turn of the century.

In 1907, for instance, the *Nation* (a New York periodical which Eliot read at Harvard) published an article on Thomson which appeared above the initials of the man whom Eliot was to call 'the finest literary critic of his time'.[18] Eliot may not have read this piece when it appeared in the *Nation*, but we can be confident that he read it when it appeared in slightly expanded form in 1908 in the fifth volume of Paul Elmer More's *Shelburne Essays*. Eliot thought highly both of their author and of this particular series of writings. Though his first published review of one of the volumes of *Shelburne Essays* is of the ninth, he there praises and reveals his familiarity with earlier volumes in the series.[19] More's article outlines Thomson's biography, his ideal love for Mathilda Weller being accorded some prominence and discussed in terms of 'these Lauras and Beatrices'. In discussing Thomson's achievement, More concentrates on a small group of poems:

. . . for it is, after all, by his four pessimistic poems—'In the Room,' 'Insomnia,' 'The City of Dreadful Night,' and 'To Our Ladies of Death'— that he has taken a unique place in literature. Some, I dare say, would reckon 'Vane's Story,' or 'Weddah and Om-El-Bonain,' or one of his two Sunday idyls as more notable pieces of writing than 'In the Room'; but there is something so singularly characteristic in this poem that it groups itself imperatively with the three acknowledged masterpieces.

[16] Lindsay (art. cit.), p. 57.
[17] Conrad Aiken, *Ushant* (1952; rpt. London: W. H. Allen, 1963), p. 71.
[18] Review of *Selected Shelburne Essays* by Paul Elmer More, *Criterion*, Jan. 1936, p. 363; see 'Paul Elmer More', *Princeton Alumni Weekly*, 5 Feb. 1937, p. 373, for Eliot's reading of *Nation*.
[19] 'An American Critic', *New Statesman*, 24 June 1916, p. 284.

After this high praise More goes on to discuss 'In the Room' in terms of Thomson's own lonely life in London lodgings, then shows the reader ' "Insomnia," with its burden of torture that impelled the poet night after night to roam the streets of London', and moves on to *The City of Dreadful Night* which is Thomson's 'phantom evocation of the London as he came to know it from his fierce nocturnal vigils'. Though More stresses that this is 'not the city of all the world', and ends by making the point that 'one is never quite permitted to escape the narrower, personal outlook in Thomson', he does quote in connection with *The City of Dreadful Night* a passage from Thomson's prose which presents a dark London as a 'vast shadowy theatre' with hallucinatory streetlamps, a city containing a fearful procession of the living and the dead.[20] It is not difficult to see how this passage, though it includes all mankind rather than a 'sad fraternity', relates to *The City of Dreadful Night*. Even the reader of that poem who is much less sensitively responsive than was Eliot can see that the relations are not only in general terms, but in matters of detail, as where the sense of not 'walking awake in a substantial city amongst real persons' corresponds to the later city which 'dissolveth in the daylight fair' (I) and of which the speaker tells us,

> I have seen phantoms there that were as men
> And men that were as phantoms flit and roam. (VII)

The words 'wherever I gaze I can discern, seeing by darkness as commonly we see by light' find their verse echo in

> And soon the eye a strange new vision learns:
> The night remains for it as dark and dense,
> Yet clearly in this darkness it discerns
> As in the daylight with its natural sense; (III)

Other correspondences could be traced, but suffice it to say that the effect of the prose passage which More quotes not only once more strengthens the link between London and the city of Thomson's poetry, but shows too how Thomson's city as London might function as a wider symbol of humanity in its horrific aspect.

Such then was the climate of opinion in which Eliot read Thomson. The older poet was associated with Dante and his City of Dreadful

[20] Paul Elmer More, 'James Thomson, ("B.V.")', *Nation* (NY), 26 Dec. 1907, pp. 583–5.

Night was seen as London. This, it should be remembered, was before
Eliot read *The Symbolist Movement in Literature* in December 1908,
where he found, for instance, the insomniac Nerval, who 'like so
many dreamers, who have all the luminous darkness of the universe
in their own brains' spent time wandering in 'uninterrupted solitude
in the crowded and more sordid streets of great cities'.[21] It was also
before he read Dante (in 1910).[22] These facts are important, for not
only did Thomson's poetry along with the English poetry of the
turn of the century provide a basis on which Eliot's reading of the
Symbolists could stand, it also, and perhaps more importantly,
shaped his reading of the great poet whose work was to be of the
highest importance to Eliot as he developed his own poetry.

Before leaving America, Eliot wrote a prayer (now in the Berg
Collection of New York Public Library) inspired by the inscription
over the gate of Dante's hell in the third canto of the *Inferno*.[23] It
is likely that the first line of Dante which Eliot read, at least in the
original, was this leading epigraph to Thomson's greatest poem, that
same terrible inscription,

> Per me si va nella città dolente

As Thomson uses it, the opening line of canto III of the *Inferno* is
designed to condition our response to 'that doleful city' (V) which
he presents to his readers and to encourage those readers to place
the details of the City of Dreadful Night with its living death within
the context of Dante's work. For Eliot, though, reading Thomson
before he read Dante, the effect must have been reversed. When he
came to read Dante, and particularly the Dante passages which
Thomson had made use of, he would have seen them in the context
of *The City of Dreadful Night* whose city, when Eliot read the poem,
was clearly linked with London. Thomson's use of the third canto
of the *Inferno* extends into his poem in the first section of which
we read the statement

> They leave all hope behind who enter there

which is an obvious use of the inscription on the Dantesque gate
mentioned in the epigraph, an inscription which Thomson later has

---

[21] Arthur Symons, *The Symbolist Movement in Literature*, intro. R. Ellmann
(New York: Dutton, 1958), p. 10; this reprints the text as Eliot read it. For the dating
of his reading, see Gordon, p. 28.

[22] A. D. Moody, p. 3.   [23] See Gordon, p. 57.

one of his city's inhabitants translate in a part of a passage which, as I shall point out shortly, contributed to the landscape of *The Waste Land*:

> I reached the portal common spirits fear,
> And read the words above it, dark yet clear.
> 'Leave hope behind, all ye who enter here:'   (VI)

There was also a subtler connection with the *Inferno* in section XVIII of Thomson's poem where, as Ian Campbell has pointed out, the creature 'crawling in the lane below' and condemned to continual slow and painful progress is like the Ser Brunetto who, condemned to move on in torture on a level below that of Dante, clutches at the skirt of the poet's coat in the fifteenth canto of the *Inferno*.[24] Other features of Dante's poem which are used by Thomson, or where Thomson and Dante at least agree, include the entering of the city from a desert. Dante enters Hell from a 'gran diserto' (I. 64), while Thomson's speaker of the famous passage with the refrain 'As I came through the desert thus it was' (IV) has entered Thomson's city from a similar locale. The idea of the circles of Dante's Hell where one of the punishments is 'roaming incessantly' (XIV. 24) contributes also to the homeless wanderings of the inhabitants of Thomson's city, and particularly to the eternal circular movement of the man who is condemned to act out

> Perpetual recurrence in the scope
> Of but three terms, dead Faith, dead Love, dead Hope.   (II)

All these features, then, relate the *Inferno* to *The City of Dreadful Night*, and the manner in which Eliot read Dante must have further encouraged him to bring to the Florentine's poem associations which had their root in the poetry of Thomson. For when Eliot came to read Dante in 1910 he did so in the traditional Harvard way outlined by Herbert Howarth.[25] This did not involve the learning of Italian grammar, but rather a direct confrontation with the original text, which did not make for fluency in Italian, since Eliot was unsure even of the pronunciation of the words and remained so all his life. Nor did it even make for an ability to understand the original on the page; Eliot read his Dante at this time, as he did later, 'only with

---

[24] Ian Campbell, '"And I Burn Too": Thomson's *City of Dreadful Night*', *Victorian Poetry*, 1978, p. 127.
[25] Howarth, pp. 72–4.

a prose translation beside the text'; but he read eagerly and, like Thomson before him, committed passages to memory.[26] Such an approach was attentive not to the rigour of fastidious scholarship and the limitations of historical conventions and philology in the accepted sense of that word, but conducive instead to an imaginative interest in the text. This method of study, bad perhaps for the scholar but good for the poet, especially the magpie and associative poet like Eliot, is a kind of reading which must have prompted him, coming to the *Inferno* via *The City of Dreadful Night*, to link the great poem with the lesser poem which had served, so to speak, as his introduction to it. Particularly in the sections of Dante which Thomson had used, Eliot would be led to associate the Florentine poet's infernal city with the 'dolent city' of Thomson's poem which was so commonly identified as London. Small wonder that Conrad Aiken noticed in London during the winter of 1921–2 when 'Eliot and myself lunched together two or three times a week in the City, near his bank' that Eliot 'always had with him his pocket edition of Dante'.[27] The *Inferno* replaced Baedeker.

The reading of Thomson, however, did not make possible for Eliot the immediate writing of a poetry dealing with the city, real or unreal. Rather, it was French poetry which let him make the necessary leap. It is Baudelaire who is clearly present beside Dante in 'The Burial of the Dead', a conjunction acknowledged by Eliot in his notes to that section of the poem. As he later put it in 'What Dante Means to Me',

I think that from Baudelaire I learned first, a precedent for the poetical possibilities, never developed by any poet writing in my own language, of the more sordid aspects of the modern metropolis, of the possibility of fusion between the sordidly realistic and the phantasmagoric, the possibility of the juxtaposition of the matter-of-fact and the fantastic.[28]

The question is how much weight can be laid on those words, 'I think', and 'developed'. Perhaps *The City of Dreadful Night* had not 'developed' such possibilities, but it had made them apparent, and Eliot had been presented in it with a London become phantasmagoric. Certainly one of the weaknesses of Thomson's poem is that, though not without certain details of real urban life, such as the

[26] CC, p. 125.
[27] Conrad Aiken, 'An Anatomy of Melancholy', in Tate, p. 194.
[28] CC, p. 126.

streetlamps whose stressed burning mixes reality with phantasmagoria to push the poem with precision towards what Paul Elmer More in the *Nation* had called its 'geometry of delirium', none the less *The City of Dreadful Night* as a whole is deficient in such concrete details drawn from the actual urban landscape.[29] Details of the landscape of the city and its environs which do emerge in Thomson's poetry are in fact incorporated into Eliot's, but it is plain that unlike the London of reality and, for that matter, of *The Waste Land*, Thomson's city is distinguished by its sparse population, whereas Eliot's surely owes more, as he himself indicated, to Baudelaire's 'Les Sept Vieillards', with its

> Fourmillante cité, cité pleine de rêves,
> Où le spectre, en plein jour, raccroche le passant!

Still, Eliot's linking of Baudelaire and Dante in *The Waste Land* and again in 'What Dante Means to Me' only serves to stress that he had read his Dante in a particular way.[30] Baudelaire was clearly the poet of a modern city, but it was Thomson who let Eliot see Dante as a poet of modern London, who let Eliot connect Dante with other modern urban writing, and who showed him that London, like Dante's infernal city, could be seen as a city of the mind as well as a city of the external world. Between 1908 and 1922 general interest in Thomson continued to surface, but it is likely that by this time Eliot had assimilated his Thomson and his memories of the poetry were germinating. Such germination was no doubt helped along by the fact that in 1917 Eliot gave as one of a 'Course of Lectures on Victorian Literature' a talk entitled 'Three Poets of Doubt—Matthew Arnold, Edward Fitzgerald, James Thomson'.[31]

The most obvious flowering to result from that germination, though not, I think, the first, comes in *The Waste Land*. Eliot's use of some Thomson material in this poem's start has been noticed above, but the most striking thing is that when Eliot comes to write of London in the poem he goes to the very cantos of the *Inferno* which Thomson had used fifty years before in his own vision of that city. So, of his crowd which flows over London Bridge (a crowd, as he

[29] More (art. cit. n. 20 above), p. 584.
[30] *CC*, p. 127.
[31] See Ronald Schuchard, 'T. S. Eliot as an Extension Lecturer, 1916–1919', *Review of English Studies*, 1974, p. 293.

later explained, of 'city clerks'—like Thomson[32]), Eliot confesses
ventriloquially, 'I had not thought death had undone so many',
showing his eye, as he, like Thomson, goes straight to Dante's
Inferno, canto III, to be on the 'prose translation beside the text' of the
Temple Classics *Inferno* with its 'I should never have believed death
had undone so many.' Eliot marked this passage in his September
1909 reprint of the Temple Classics edition.[33] Eliot next goes to his
Dante for the line, 'Sighs, short and infrequent, were exhaled', in
connection with which his note refers the reader to canto IV of the
Inferno:

> Quivi, secondo che per ascoltare
> non avea pianto, ma' che di sospiri,
> che l'aura eterna facevan tremare.   (ll. 25–7)

of which the Temple Classics translation is 'Here there was no plaint,
that could be heard, except of sighs, which caused the eternal air
to tremble'; and this too is a passage which with its suggestion of
suppressed grief and of human breathing affecting the whole air is
related to a Thomson motif which recurs in *The City of Dreadful
Night* and which is found, for instance, in section XV of that poem,

> Wherever men are gathered, all the air
> Is charged with human feeling, human thought . . .
> Unspoken passion, wordless meditation,
> Are breathed into it with our respiration . . .

No doubt Thomson himself had been attracted to the Dante passage
by the condition of those prisoners of the *Inferno* who are described
as desiring without hope.

Again, Eliot's final use of Dante in this section of *The Waste Land*,
in the recognition scene, 'There I saw one I knew, and stopped him,
crying: "Stetson!"' seems to make use of the recognition scene in
the Ser Brunetto passage of the *Inferno*, once more a passage used,
as we have seen, by Thomson. Eliot marked *Inferno* XV. 16–21 in
his Temple Classics copy.

In *The Waste Land* the London Bridge passage fuses Baudelaire
and Thomson's Dante. London becomes both Baudelaire's city of
the daylight ghost, but also in its Dantesque gloom (Eliot marked

---

[32] *CC*, p. 128.
[33] John J. Soldo, 'Eliot's Dantean Vision, and his Markings in his Copy of the
Divina Comedia', *Yeats Eliot Review*, 1982, p. 17.

Inferno's 'e l'aere bruno' in his copy), 'Under the brown fog of a winter dawn', it becomes the Thomson city of death-in-life where men and phantoms are indistinguishable from one another, the 'Unreal City' in the words which introduce the passage, and the city, too, where, a few lines before, Madame Sosostris senses for the inhabitants the fate meted out to Thomson's celebrated doomed individual, 'I see crowds of people, walking round in a ring.'

But from Thomson Eliot did not only learn how to see Dante's *Inferno* as being relevant to modern London, for Thomson also provided examples of the squalid urban settings which Eliot was to use in *The Waste Land* and showed how apparently unpoetic urban sights could be made to communicate powerfully a sense of horror. In section VI of *The City of Dreadful Night*, the section in which the literal translation of the inscription over the infernal gate occurs, a figure is sitting 'forlornly by the river-side' where he hears in a 'waking dream' the sound of 'bodiless voices' who complain about the vacuity of life, 'this insufferable inane'. This same setting is used by Thomson in section VIII of his poem where, at the riverside, beside 'the tide as black as our black doom' another couple are overheard in conversation,

> 'We gaze upon the river, and we note
> The various vessels large and small that float,
> Ignoring every wrecked and sunken boat.'

Eliot's city too has its river with its debris and its water black where 'The river sweats / Oil and tar'. Thomson's use of an unpleasantly damp atmosphere as he walks on 'earthen footpath, brushing as I went / The humid leafage' (XVIII) is matched, when married with Thomson's canal bank, by Eliot's riverbank where 'the last fingers of leaf / Clutch and sink into the wet bank'. The ensuing 'dull canal' of *The Waste Land* is a feature too of Thomson's squalid poetic landscape in 'A Real Vision of Sin' where by a 'foul canal' an old man and woman, 'two old bags of carious bone', discuss their past, ruined love in a landscape presented in terms of damp horror after the striking opening,

> Like a soaking blanket overhead
> Spongy and lax the sky was spread,
> Opaque as the eye of a fish long dead.

We are shown under a landscape of 'drizzling weather' the river with

> a slushy hollow near its bank,
> Where noisome weeds grow thick and dank.

This world, a place of 'green scum' rotting, a landscape of the 'water-rat', is one in which the man, 'red-rat-eyes', argues with and is finally the killer of the hag, that old bony creature's reviling playing off against the remembered erotic details:

> This dirty crumpled rag of a breast
> Was globed with milk once; I possest
> The means of being grandly blest!
>
> 'Did the babe of mine suck luscious sips,
> Soothing the nipple with rose-soft lips
> While her eyes dropped mild in a dear eclipse? . . . '

The same landscape and details are condensed in Eliot's

> A rat crept softly through the vegetation
> Dragging its slimy belly on the bank
> While I was fishing in the dull canal

and here too we have the uneasy potency of the combination of erotic detail in this unhealthy setting, 'White bodies naked on the low damp ground'. If the 'crumpled rag of a breast' in this poem suggests the 'wrinkled dugs' of the Tiresias found by Eliot in *Inferno* XX and in other nineteenth-century poetry, then it should be noted too that the daringly 'unpoetic' vocabulary of this striking poem shows Thomson also anticipating Eliot in his use of an adjective most unusual in verse, namely 'carious', which is found in Eliot's 'Dead mountain mouth of carious teeth that cannot spit'.

On a structural level, contained in Thomson's *City of Dreadful Night* is an account of travelling through the desert, as we find in Eliot's *Waste Land*. In both cases the combination of city and desert is the legacy of the romantic view of the city as a desert, Wordsworth's 'wilderness of building'.[34] Thomson's city is a 'builded desolation' (V) as spiritually desolate as Eliot's London. But if in Thomson's city 'There is no God' (XIV), Eliot's view is not so clear-cut. In this respect and in some others Eliot's poem is closer to Thomson's earlier 'The Doom of a City' which, like the original *Waste Land*, contains a sea voyage, but, more importantly, has a third section, 'The Judgments',

---

[34] Wordsworth, *Poetical Works*, ed. J. O. Hayden (Harmondsworth: Penguin, 1977; 2 vols.), ii. 90 (*Excursion*, II. 836).

in which, while the inhabitants pray 'Grant us awhile Thy perfect peace, O Lord', a 'great Voice' speaking in thunder from a thunderstorm pronounces a destructive judgement on the city below. Though far lengthier than what Eliot's thunder said, this judgement is expressed in block capitals, the typographical dress of Eliot's thunder voice, and after the judgements the Thomsonian narrator returns home in his boat bringing a warning to his own city which is very like that described in *The City of Dreadful Night*, except that this 'Empress in thine own right of the earth-surrounding sea' is even more obviously London. Thomson's influence, then, need not be confined to the details of Eliot's poetry, but may be present too in the structure of *The Waste Land*. Eliot's use of the thunder voice in *The Waste Land* comes directly from his Sanskrit studies and is encouraged by his anthropological reading, but again this material overlies elements of Victorian romanticism which it revives.

Eliot's reading of Thomson did not affect only *The Waste Land*. 'Prufrock' is very much a poem influenced by his knowledge of French poetry, but it is also an ironic view of the type of figure found in 'Insomnia', a Thomson poem written less than thirty years earlier. In the Dr Sweany world of 'Insomnia' we see the familiar sleepless figure wandering the city streets self-consciously aware of his own impotence:

> When some stray workmen, half-asleep but lusty,
> Passed urgent through the rainpour wild and gusty,
> I felt a ghost already, planted watching there.

Aware of the fruitlessness of his own year, this figure has none of the wryly self-conscious and nervously ironic amusement of Eliot's figure. The persona of an unpublished poem, feeling himself (when attending *Tristan and Isolde* in 1909) to be like the ghost of youth at the undertaker's ball, is more complex, funnier, and, for that reason, potentially lively.[35] Yet Thomson's figure has, like Prufrock, a self-consciousness of the insufficiency not only of his own actions but of his own words as, like the later nocturnal wanderer, he tails off into the dot dot dot of paralysis:

> I look back on the words already written,
> And writhe by cold rage stung, by self-scorn smitten,
> They are so weak and vain and infinitely inane. . . .

[35] 'Opera' in The Complete Poems MS (Berg).

As Prufrock is conscious, thanks to Laforgue no doubt, that 'No! I am not Prince Hamlet, nor was meant to be,' so Thomson elsewhere tells us that 'whosoe'er the destined roles may fill, / Hamlet is Hamlet — Osric Osric still';[36] but more noticeably in 'Insomnia' we again find a Dantesque reference which seems to have guided Eliot's use of Dante. In the lines which follow the last passage quoted from 'Insomnia', Thomson presents his readers with an apparent quotation from Dante,

> 'How from those hideous Malebolges deep
> I ever could win back to upper earth,
> Restored to human nights of blessed sleep
> And healthy waking with the new day's birth?'

The lines are not a quotation from Dante, yet since the poem 'Insomnia' has but one speaker throughout, Thomson's placing of these words within quotation marks makes them appear to be a Dantesque borrowing, all the more so since the word 'Malebolge' is first cited by the *OED* as occurring in 1861 which would make it in this poem of 1882 an even more obviously Dantesque marker than it is today. It is, I suggest, this placing of the wanderer of the modern city streets in the depths of the Inferno which guides Eliot's placing of his own later nocturnal wanderer, by means of the epigraph to 'Prufrock' in the same region. Thomson died shortly after the writing of 'Insomnia', as Eliot must have known.

The fear of inanity, a constant fear in Thomson's poetry, was to be a crucial theme in the work of Eliot. Longing for life, for experience to fill emptiness, and a sense of the horror of 'Death-in-Life' (*City of Dreadful Night*, V) are seen, for instance, in lines from an early Thomson poem quoted by Dobell in his 'Memoir' prefaced to the two volume *Poetical Works* in which, trapped in 'this lone base flat of torpid life', the speaker longs for action, complaining,

> I fret 'neath gnat-stings, an ignoble prey,
> While others with a sword-hilt in their grasp
> Have warm rich blood to feed their latest gasp.[37]

This, refocused, becomes the complaint of Gerontion who has not

[36] Thomson, *PW*, i. 275 ('Prologue to the Pilgrimage of Saint Nicotine of The Holy Herb').
[37] Ibid., p. xxxvi ('Twenty-third Birthday').

> fought in the warm rain
> Nor knee deep in the salt marsh, heaving a cutlass,
> Bitten by flies, fought.

But, as in Thomson, the horror of inanity is accompanied by a sense of repressed passion such as that of 'In the Room' when 'throughout the twilight hour . . . Life throbbing held its throbs supprest.' So in Eliot's London waste land,

> At the violet hour . . .
> . . . the human engine waits
> Like a taxi throbbing waiting, . . .

For both poets life could seem a desert, and London an infernal region; the later was to find a way out of both which the earlier writer was never able to discover or, perhaps more accurately, to rediscover. It is not surprising that Eliot's reading of this nineteenth-century poet-clerk should have maintained an effect on his own writing. Eliot later worked, like Thomson, in a London office. The two men shared a concern with the city seen as a locus for despair, as well as a search for a peace passing understanding; both writers had the experience of an uprbinging filled with Calvinist morality. Thomson's effect on Eliot is certainly complicated by being overlaid by the influence of other poets justly and often seen as crucial guides in the formation and development of his style and matter; but Thomson is concealed as a constituent element in the 'familiar compound ghost' of Eliot's poetic ancestry. It is but another indication of Thomson's influence that when Eliot talks of this ghost in section two of 'Little Gidding' he goes once more to a Dante passage used in *The City of Dreadful Night*, just as, when speaking of the doomed businessmen of modern London in part three of 'East Coker' with its 'O dark dark dark. They all go into the dark', Eliot had made use of a Milton tag seen first in *The Leavenworth Case*, but used too by Thomson of the inhabitants of his city:

> O melancholy Brothers, dark, dark, dark! . . .
> O dark, dark, dark, withdrawn from joy and light! (XIV)

When the 'familiar compound ghost' is met in 'Little Gidding', it is clear not only that Eliot is making use of the Ser Brunetto passage in the *Inferno* once more, but also that the figure met

> In the uncertain hour before the morning
> Near the ending of interminable night

is a figure like the insomniac Thomson stravaging the apparently
interminable streets of a dreadful nocturnal London. Where Thomson
had met his Ser Brunetto figure at 'a spot whence three close lanes
led down' (XVIII), Eliot's meeting takes place, as Grover Smith has
pointed out, in a similarly trisected spot where on

> the asphalt where no other sound was
> Between three districts whence the smoke arose
> I met one walking, loitering and hurried.[38]

This man with 'the down-turned face', given the way in which
Thomson's landscape became incorporated into Eliot's, and the way,
more importantly, in which the reading of Thomson affected Eliot's
reading of Dante, now seems compounded out of Thomson as well
as out of Dante and Eliot's other masters. Anne Ridler's suspicion
in 1963 that 'Thomson might even have some part in the "familiar
compound ghost" of "Little Gidding"' was well founded and can now
be justified.[39] No longer is it necessary to write, as Ian Campbell
did, that 'curiously enough' both Thomson and Eliot make use of
the same Ser Brunetto passage.

   Thomson, like Eliot and Dante, was a man who spoke 'From
the midst of the fire',[40] feeling 'The Judgments' of 'The Doom of
A City' 'A torture of intolerable fire', just as Eliot was to know 'The
intolerable shirt of flame' which was finally incorporated into the
fiercely striven-for vision of 'Little Gidding'. Thomson is a lesser artist
than either of the two other poets, but he is one who contributed
both to Eliot's mental landscape and to the landscape, most notably
that of the 'Unreal City', which Eliot places before the reader of his
poetry.

   That 'Unreal City' is so striking partly because it conforms to our
normal notions of reality. We can find London Bridge and King
William Street on the map. Eliot's philosophical studies and Buddhist
interests show him deeply fascinated by the elusive nature of reality.
His work on Bradley led him towards the borders of what we
normally term the real. He was interested in three-dimensional
hallucinations and visions, writing in 1913 a detailed excursus on

[38] Grover Smith (1974), p. 342 n. 95.
[39] Anne Ridler, ed., *Poems and Some Letters of James Thomson* (London:
Centaur Press, 1963), Intro., p. xliii.
[40] From Thomson's lines included in his 'Introductory Note' to 'A Lady of
Sorrow', *Essays and Phantasies*, p. 1.

someone who is convinced he sees a bear, while the creature is perceived by no one else.[41] In his work on reality/unreality Eliot refers, as he would several times in published prose, to the work of Pierre Janet on hallucination among the hysterical, where subjects experienced 'the sensation of touch and weight as well'.[42] In 1916 Eliot concluded that 'In order to acknowledge the existence of hallucinations we have partially to concede their truth.' Reality appeared an arbitrary construct. 'We only say that an object is unreal with respect to something else which we declare to have been affirmed at the same time and which continues to be real, while the other does not.'[43] *The City of Dreadful Night*, at once seen as Thomson's London and as a phantasmagoria, fitted well with the philosophical interests which Eliot would develop and with his own 'Unreal City'.

But, as in St Louis so at Harvard, Eliot knew an actual city, and enjoyed several of its distinctly urban pleasures. As a Harvard undergraduate he enjoyed melodrama at Boston's Grand Opera House.[44] It was a taste which would lead him to Marie Lloyd, and it was a taste common amongst his fellows, one of whom saw the typical last act of a comic opera as being set in 'a street in the Big City. On the right, palatial residence of Jacob Porous Plaster—on the left, a horrid slum.'[45] Such confrontations of the genteel and the slummy could be seen in Boston and in Eliot's early poetry. There is gloom and terror in some of his early unpublished city poetry. There is a preoccupation with debris and a shattered world, but there is also a humour that sets details of vulgar life against a romantic, genteel world of blue-delft china. In 'Goldfish (Essence of Summer Magazines)', the ladies embarking for Cythera take ginger beer and sandwiches.[46] The songs of the contemporary comic opera should not be dismissed as lacking in intellectual content, or merely sentimental. Often they made fun of 'higher' learning, or could be adapted to do so.

[41] 'Degrees of Reality', MS, 1913 (King's College Library), pp. 4–5. This passage was revised for Eliot's thesis. See *KE*, pp. 115–16.

[42] *KE*, p. 115; cp., e.g., 'Janet was the great psychologist [in Paris around 1910]'—'A Commentary', *Criterion*, Apr. 1934, p. 452.

[43] *KE*, p. 115.   [44] *Facsimile*, p. 125 n. 6.

[45] H. P., 'The Influence of the Comic Opera', *Harvard Advocate*, 83 (1908), p. 99.

[46] The Complete Poems MS (Berg), p. 19 ('Embarquement pour Cythère').

She learnt, goodness gracious, that God is fallacious—
A theory that Boston reveres.
But she grew pretty sour on old Schopenhauer
And Nietsche [*sic*] reduced her to tears.[47]

In Eliot's 'Suite Clownesque', written around the time of 'Prufrock'
and 'Portrait of a Lady' a red-nosed clown (a comic Cyrano?)
philosophizes among potted palms and serenades, a chorus of seven
little emancipated maids sings about flirting and construing texts,
and the suite ends with a song about Broadway after dark which
alludes to several musicals as well as to philosophical themes. A stage
direction asks for a clownesque to be played using bones and
sandboard.[48] The jazz rhythms of this lively, cynical piece reinforce
the idea that it was the result of a fascination with musicals which
would later present 'SWARTS AS TAMBO. SNOW AS BONES', after
including several references to popular songs from *The Waste Land*
manuscripts.[49]

These curious juxtapositions seen in Eliot's early poems owe something to Laforgue, a debt about which Eliot and others have written
at length.[50] But the world of the music hall was also important.
Music hall was very much a feature of 1890s poetry. Eliot, however,
is close not so much to Arthur Symons as to another 1890s poet
whose suicide in 1909 made him especially topical during Eliot's
Harvard years, and whose first book of verse was entitled *In a Music
Hall and Other Poems*. John Davidson wrote bitter poems to popular
tunes such as 'After the Ball'. In his work the everyday and vulgar
continually meets the terrors of the infinite in an imagination, like
Eliot's, both apocalyptic and deadpan. Haunted by the waltz of a
street piano, Davidson's speaker wonders,

Grace and delight and beauty and youth,
Will they go out like the lights at a ball
With sun, moon and stars, at the end of all?[51]

[47] H. P. (art. cit.), p. 99.
[48] The three sections of 'Suite Clownesque' appear on pp. 24–7 of the Complete
Poems MS (Berg).
[49] *CPP*, p. 122; *Facsimile*, p. 125 nn. 2, 3, 5, 7; *The Waste Land* kept, of course,
its Australian ballad and 'Shakespeherian Rag'.
[50] See esp. E. J. H. Greene, *T. S. Eliot et la France* (Paris: Boivin, 1951).
[51] John Davidson, 'To the Street Piano, II., After the Ball' (from *Ballads and
Songs*, 1894), in *The Poems*, ed. Andrew Turnbull (Edinburgh: Scottish Academic
Press, 1973, 2 vols.), p. 63.

In the draft of 'Prufrock' Eliot also ended the world.[52] If James Thomson introduced Eliot to Dante, then John Davidson, a poet preoccupied with the 'masterdom' of 'self-consciousness' played a crucial role in introducing him to Laforgue and the French symbolists, as Eliot himself acknowledged.[53] Eliot read Davidson around the same time as he read Thomson, probably a little afterwards, and tended to connect the two, while recognizing their differences.[54] The most crucial of these differences is the sharpness of Davidson's eye. 'The truth', wrote Eliot in 1917, 'is that Davidson was a violent Scotch preacher with an occasional passionate flash of exact vision.'[55] Both the passion and the exactness were important to Eliot. Davidson also had his ghostly city, a place, like Eliot's London, where phantoms move among actual street names. The world of real and unreal, of dead and living can seem as uncertain in Davidson's poetry as they can in *The Waste Land* or *Sweeney Agonistes*

> I move from eastern wretchedness
>   Through Fleet Street and the Strand;
> And as the pleasant people press
>   I touch them softly with my hand,
> Perhaps to know that still I go
>   Alive about a living land.[56]

Baudelaire's 'spectre' who 'en plein jour raccroche le passant' was already in London. Eliot read widely in Davidson, admiring his lyrics, ballads, and 'A Runnable Stag'.[57] In 'A Loafer', for instance, Davidson could shift rapidly from city streets to ocean depths, 'where sea-whelmed the drowned folks lie'. The loafer about whose 'face like seaweed droops / My tangled beard, my tangled hair' sent Eliot looking 'for the head of Mr. Apollinax . . . With seaweed in its hair', while Davidson's 'Winter Rain', as Grover Smith points out, would find its way into 'Lines to a Yorkshire Terrier'.[58] Eliot thought one critic 'probably right in referring to the *Ballad of a Nun* as "an outburst of hysteria"', but religious hysteria, whether in the pages

[52] Complete Poems MS (Berg), pp. 33–4.
[53] Davidson, *The Testament of a Prime Minister* (first published in 1904), *The Poems*, p. 372; Eliot cited in M. Lindsay (art. cit.), p. 57.
[54] 'Preface' to *Davidson* (B83), pp. xi–xii.
[55] 'Reflections on Contemporary Poetry [II]', *Egoist*, Oct. 1917, p. 134.
[56] Davidson, 'A Loafer' (from *Selected Poems*, 1905), *The Poems*, p. 80.
[57] See n. 54 above.
[58] 'A Loafer', see n. 56 above; Grover Smith (1974), p. 254.

of Janet or in his own early poems, clearly interested Eliot.[59] Davidson's vocabulary was exciting. Poet of a 'waste raw land', he could use a word like 'protoplasm' in verse published over a decade before 'Burbank with a Baedeker'.[60] But most of all, he mattered as a poet of the city. When Eliot's 'crowd flowed over London Bridge', a crowd he later described as 'city clerks trooping over London Bridge from the railway station to their offices', he was in the territory of Davidson's 'A Certain City Terminus', where London Bridge station

> Discharges into London's sea, immense
> And turbulent, a brimming human flood

of workers who move in 'dull, rapid fashion' with impersonal, despairing faces

> Oblivious, knowing nought terrestrial
> Except that time is money, and money all.[61]

Eliot had 'a fellow feeling with the poet who could look with a poet's eye on the Isle of Dogs and Millwall Dock'.[62] Davidson described the polluted Thames 'Among the barges'.[63] He wrote of how hearing 'The Old Hundredth tune' made momentarily a paradisal 'northern isle' out of 'The Isle of Dogs' in its landscape of boats anchored by shabby rented houses.[64] In *The Waste Land* Eliot would set the Rhine maidens' song against the polluted Thames where

> The barges wash
> Drifting logs
> Down Greenwich reach
> Past the Isle of Dogs.

With Thomson, Davidson consecrated such material for Eliot's poetic use. Though he did not wish 'to give the impression that for me

---

[59] See n. 55 above.

[60] Davidson, 'A Woman and her Son' (from *New Ballads*, 1897), *The Poems*, p. 303; *The Testament of a Prime Minister* (1904), *The Poems*, p. 371.

[61] This poem appeared as 'A Certain City Terminus' in Davidson's *Fleet Street and Other Poems* (1909); rpt. as 'Railway Stations, 1. London Bridge', *The Poems*, pp. 434–6; cp. *CC*, p. 128.

[62] 'Preface' to *Davidson* (B83), p. xii.

[63] *The Testament of a Prime Minister* (1904), *The Poems*, p. 355.

[64] Davidson, 'In The Isle of Dogs' (from *The Last Ballad and Other Poems*, 1899), *The Poems*, pp. 133–4.

Davidson is the author of one poem only', it is undeniable that one particular Davidson poem affected Eliot most deeply.[65] 'Thirty Bob a Week' was a poem which his Harvard friend W. G. Tinckom-Fernandez would recite and which Eliot would call 'great',[66] finding in it 'inspiration' in its revealing of dingy urban images and its marrying such subject-matter to diction not conventionally poetic. 'The personage that Davidson created in this poem has haunted me all my life, and the poem is to me a great poem for ever.' These are strong words to use, particularly for Eliot, and make it all the more surprising that Davidson's work has been so little discussed in relation to Eliot's. Clearly Eliot was attracted to the way in which in this poem 'Davidson freed himself completely from the poetic diction of English verse of his time.'[67] Davidson's style, buttonholing and antipoetic in some ways, allows a sharper presentation of thought and of emotional depths free from the fug of late romantic diction.

> I ain't blaspheming, Mr. Silver-tongue;
>   I'm saying things a bit beyond your art:
> Of all the rummy starts you ever sprung,
>   Thirty bob a week's the rummest start!
> With your science and your books and your the'ries about spooks
>   Did you ever hear of looking in your heart?[68]

The figure of the clerk is a common one in Davidson's poetry. 'In a Music-Hall' begins

> In Glasgow, in 'Eighty-four,
> I worked as a junior clerk;
> My masters I never could please,
> But they tried me a while at the desk.

This clerk, working in the office day after day, loses his soul nightly in the music hall, and his attitude has in it the spirit that would attract Eliot to Baudelaire, a wish to escape from an inane world.

> It is better to lose one's soul,
> Than never to stake it at all.[69]

---

[65] 'Preface' to *Davidson* (B83), p. xii.

[66] Aiken, *Ushant*, p. 133; *SE*, p. 336 n.

[67] 'Preface' to *Davidson* (B83), pp. xi–xii.

[68] Davidson, 'Thirty Bob a Week' (from *Ballads and Songs*, 1894), *The Poems*, p. 64.

[69] Davidson, 'In a Music-Hall' (from *In a Music-Hall*, 1891), *The Poems*, pp. 21–2.

In 'Thirty Bob a Week' Davidson's clerk, by stepping into his heart
enters a world of violence beneath the surface of life. Such hidden
depths beneath the clerkly life, familiar to Eliot from Sherlock
Holmes stories and *The Leavenworth Case*, as well as from Marlow
in *Heart of Darkness* and some of Kipling's London stories, would
be central to his own work, underpinning the middle poetry of *The
Waste Land* and *Sweeney Agonistes*. These depths are present also
in the early unpublished poems where the surface of city life breaks
open to reveal a powerful life beneath, or where small dusty souls
fear divine scrutiny. Looking directly inwards for the dweller among
the numbing movements and the unpleasant, yet routine debris of
urban life is a business evaded ('O do not ask what is it') because
shocking. The monotony of living wears down the clerk in his rented
rooms, but it avoids the gaze to the heart of things.

> For like a mole I journey in the dark,
>   A-travelling along the underground
> From my Pillar'd Halls and broad Suburbean Park,
>   To come the daily dull official round;
> And home again at night with my pipe all alight,
>   A-scheming how to count ten bob a pound.

Here again, Davidson consecrated 'unpoetic' landscapes of pipe-
smoking suburbs and the Underground for the Eliot who would
himself be a hard-pressed bank clerk with financial worries and who
would make poetry out of the London subway in *The Rock* and in
Four Quartets. In 'East Coker' the Underground, like Davidson's,
is associated with the routine of life under which a terror (associated
with inanity) may lurk

> . . . as, when an underground train, in the tube, stops
>   too long between stations
> And the conversation rises and slowly fades into silence
> And you see behind every face the mental emptiness deepen
> Leaving only the growing terror of nothing to think about.

By *Four Quartets*, though, such an inner journey is not only frightening; it also holds the possibility of an eventual light through
the darkness, a hope beyond hope. Eliot was preoccupied in his
earlier writing with the clerk's perception only of depths of 'The
horror! the horror!'[70] Davidson's clerk in 'Thirty Bob a Week'

---

[70] Eliot quoted this phrase from *Heart of Darkness* as an epigraph to the MS of
The Waste Land; see *Facsimile*, p. 3.

perceives in his city heart something as horrible as that observed by
Marlow in the jungle:

> I step into my heart and there I meet
>   A god-almighty devil singing small,
> Who would like to shout and whistle in the street,
>   And squelch the passers flat against the wall;
> If the whole world was a cake he had the power to take,
>   He would take it, ask for more, and eat them all.

Davidson's poem gains in force from the way a directly colloquial
diction is used at once with precision and with anger. None of Eliot's
personae gets that angry; their poet is more urbanely humorous, and
only Mr Apollinax a devourer in a world of bitten macaroons with
'His laughter . . . submarine and profound' gets as deep. But if he
is unable to wield such violent feeling, clerkly Prufrock is its victim.
Unlike the clerk of 'Thirty Bob a Week', he does not 'squelch the
passers flat against the wall' but he does see himself, a martyr to
smooth words, 'pinned and wriggling on the wall'. In general example
and in verbal detail, Davidson contributed much to the Prufrock who
garrulously seeks a way of putting things, at once concerned to reveal
his soul yet terrified and unable to do so, talking of spitting, biting,
squeezing, and ragged claws, as Davidson's clerk had pictured himself
as a 'mollusc'. The earlier poet's clerk falls 'face forward, fighting,
on the deck', but Prufrock retreats to *his* end, timorously unable to
live up to his more aggressive counterpart's challenge, 'You say it
if you durst!'

The powerful emotion of 'Thirty Bob a Week' is all the stronger
for its factual quality: its stress on cash and specific urban images
such as rented rooms 'the size of travelling trunks' or 'the god-
almighty devil and the fool / That meet me in the High Street'.
The factual precision strengthens the emotion. The sigh is more
complex and more powerful than a meaningless romantic sigh
when 'we cough, my wife and I, to dislocate a sigh'. As the clerk
puts it,

> And I've got to cut my meaning rather fine,
>   Or I'd blubber, for I'm made of greens and sod.

Eliot employs a similarly aggressive mixture of urban hardness and
tender feeling in 'Preludes'.

> I am moved by fancies that are curled
> Around these images, and cling:
> The notion of some infinitely gentle
> Infinitely suffering thing.
>
> Wipe your hand across your mouth, and laugh;
> The worlds revolve like ancient women
> Gathering fuel in vacant lots.

The danger of pathos lapsing into sentimentality does not occur in 'Thirty Bob a Week', being warded off by a proud violence. Eliot's use of such a violence in his first book is not always successful. 'The last twist of the knife' concluding 'Rhapsody on a Windy Night' topples bitter irony into melodrama. Eliot's tendency to go over the top (which Pound curbed in the case of the 'house agent's clerk' of *The Waste Land*[71]) also leads in 'Portrait of a Lady' to the too obvious repetition of

> And twists one in her fingers while she talks. . . .
> (Slowly twisting the lilac stalks)
> 'You let it flow from you, you let it flow, . . . '

It is necessary to remember the concern with impotence, languor, and stultifying amoral gentility in Eliot's early poems, but also to remember their subversive humour and violence. The humour is largely Eliot's own. In the deserts of Fitzgerald's *Rubáiyát* and Thomson's London he found the languor and impotence he would present in his city poems; in Davidson's London and his city clerks he located both violence and exactness of diction. Clearly Eliot found in these writers many other elements that filtered through his life into his poetry. Laforgue, Baudelaire, and the Symbolists were the catalyst which sparked off Eliot's own voice. But they activated material which he had read earlier. When that Eliotean voice presented its city and that city's inhabitants, the gloomy language of late nineteenth-century English poetry was dislocated into a sharper meaning.

[71] *Facsimile*, p. 47.

III

SAVAGE

ABOUT fifteen months after lamenting in 'The Man Who Was King' that the natives of Matahiva were now 'quite civilized and uninteresting', Eliot was at Harvard, and 'savages', to use the terminology which he found in the anthropology of the day, were popular.[1] In 1907 the *Harvard Advocate* published a story whose white protagonist is directed into the jungle by 'Buldino tribesmen' and eaten alive by ants.[2] Blended with the crucifixion meted out by the tribesmen of Kipling's 'The Man Who Would be King', this fate would emerge as Celia's in *The Cocktail Party*. If particular details impressed Eliot, however, they were only a part of the general atmosphere of the time.

'PRIMITIVE MAN' read the large type heading to the large type account of the leading article when *Putnam's Monthly* advertised itself in 1907 as carrying accounts of discoveries of skulls suggesting 'the existence of a race of inferior intelligence to any other of which records exist'.[3] The primitive was in vogue. The 1908 *Nation* carried a report showing how Arnold's 'Forsaken Merman' (to be used later by Eliot in 'Prufrock') represents 'the crudeness and even brutality of . . . ancient legend refined and sublimated'.[4] There was a popular journalistic awareness of how 'survivals of savage passion serve to remind us how thin is the veneer of our civilization, how easy it is to drop back to the moral level of the ape and tiger'.[5] In 1912 the bicentenary of Rousseau quickened further interest in notions of primitivism, so that Irving Babbitt thought Bergson and Rousseau 'perhaps the two men most talked and written about internationally of late', Bergson himself being 'plainly a Rousseauistic primitivist'.[6] While the study of primitive music, dancing, and

---

[1] 'The Man Who Was King' (C4), p. 1.
[2] L. D. Cox, 'The Creeping Gods', *Harvard Advocate*, 83 (1907), pp. 27–30.
[3] '*Putnam's Monthly* for 1907' (advertisement), *Nation*, 3 Jan. 1907, front cover.
[4] 'The Forsaken Merman', unsigned article, *Nation*, 9 Jan. 1908, p. 29.
[5] 'The Mystery of Cruelty', unsigned article, *Nation*, 28 May 1908, p. 483.
[6] Irving Babbitt, 'Bergson and Rousseau', *Nation*, 14 Nov. 1912, pp. 452 and 453.

literature was being pressed forward, so was the study of the development of ritualized sexuality ranging 'from primitive anthropology to modern—and decadent—civilization'.[7]

During Eliot's time at Harvard, modern urban man and primitive man were often juxtaposed; the expansion of Darwinian thought and of modern anthropology were part of a widespread concern with the inheritance of genetic characteristics and the transmission of cultural values. For many anthropological writers, primitive man was at one end of the scale, modern western man at the other. But there was a fascination with connecting the two, not only in the work of scholars, but in popular culture and in the feats of travel and exploration common at the turn of the century when Eliot's sensibility was forming. *The Waste Land* was to draw on the experience of Shackleton (records of whose expeditions were being published in Eliot's student days), and had an iceberg seascape in a cancelled section. These metaphors for humanity *in extremis* reflect the journeys of the travellers who were newsworthy in Eliot's student years. Later too his eye continued to fall, for instance, on a Victorian sculptor, whose letters described his adventures as a seeker for gold in the Australian bush, or, among a catalogue of mediocre First World War poets, on Brian Brooke, who was particularly interesting to Eliot since, as a settler in East Africa, he had been called Kurongo (strong man) by his neighbours and adopted into their tribe. 'He was a genuine explorer in temperament, having (as we infer) that particular matter-of-fact romanticism which is proper to explorers and adventurers.'[8] In Eliot's youth, materials such as these were increasingly being channelled towards the growing study of anthropology.

What constituted anthropology, 'this most unsettled and vaguely limited field of study', was uncertain.[9] Yet this also had the effect of involving anthropological material in some of the most controversial issues of the day, such as Mendelian biology, eugenics, sociological investigation, and the development of the study of religion. Proof that this academic subject was publicly noteworthy is provided by its being used for satirical purposes. In 1914, parodying

[7] Unsigned review of Friedrich S. Krauss, *Anthropophyteia*, vol. iii and *Der Volksmund*, 'Science', *Nation*, 21 Mar. 1907, p. 272.

[8] 'A Victorian Sculptor', *New Statesman*, 2 Mar. 1918, p. 529; 'The New Elizabethans and the Old', *Athenaeum*, 4 Apr. 1919, p. 135.

[9] 'Science', unsigned article, *Nation*, 2 July 1908, p. 18.

works such as Westermarck's 1906 *History of Human Marriage*, the *Nation* printed under the heading 'The Complete Anthropologist' a mock review of an 'epoch-making book on "Modern Woman and Marriage by Capture"' in which one 'Professor Junker of the University of Zabern has demonstrated in masterly fashion how every physical and mental trait of the woman of to-day may be traced back to the time when the primitive male, whenever he wanted a wife, went out and caught one'.[10] If this and other fun-poking assumes a popular knowledge of the sort of material anthropologists produced, it also hints at why anthropology gained its public notice. This was because it was closely concerned with what had been the most unsettling topic to confront the past few generations: the study of origins. In 1909, the fiftieth anniversary of the publication of *The Origin of Species*, both the *Dial* and the *Nation* spent some time reviewing *Fifty Years of Darwinism* which included contributions by 'Schwalbe, one of the most eminent of living anthropologists' (a critical piece on *The Descent of Man*), Haeckel ('Charles Darwin as an Anthropologist'), and Frazer ('Some Primitive Theories of the Origin of Man'). The same book contained work on heredity and genetics by scientists including De Vries, Bateson, and Davenport, alongside what the *Dial* saw as 'the most interesting part of the book', that dealing with 'so-called "humanistic" fields of thought' which showed 'how tremendous has been the influence of Darwin in fields far removed from those in which he himself worked'.[11] These essays included Jane Harrison on 'The Influence of Darwinism on the Study of Religions'. Paul Elmer More wrote the next year of how the Victorian public had sensed that *The Origin of Species* had been the book which 'had struck the keynote of the [nineteenth] century', encouraging what he called, with a possible side-glance at the modern obsession with Bergson, 'the philosophy of change'.[12] Through the notion of evolution anthropology was linked with numerous other areas of study. In its particular concern at this time with the changes which had produced 'civilized' humanity and the understanding of which might show how sophisticated modern man was linked to as well as distanced from primitives and savages, anthropology continued the sort of probings and generated the sort of unease which

[10] 'The Complete Anthropologist', unsigned skit, *Nation*, 2 Apr. 1914, p. 355.

[11] Raymond Pearl, 'Taking Stock of Darwinism', *Dial*, 16 Aug. 1909, p. 93.

[12] Paul Elmer More, 'Victorian Literature (The Philosophy of Change)', *Nation*, 6 Oct. 1910, p. 309.

Darwin's work had sparked off fifty years earlier. Sir E. B. Tylor
and his fellows had 'set the world right in regard to the actual
beginnings and early progress of human society'.[13] For its treatment
of 'social evolution' W. G. Sumner's *Folkways* was hailed in 1908
in words pleasing to its author as 'a book which, it seems to the
writer, Charles Darwin would have written if he could'.[14]

From disciplines as various as physiology, philosophy, economics,
and neurology, anthropologists arose and the new subject was
recruited under various banners by men such as Sir Francis Galton,
explorer and (now much criticized) anthropologist and pioneer of
'eugenics'. 'Galton,' we hear in 1912, 'by his researches in heredity,
is regarded as having initiated the modern study of race improvement';
under his influence some psychologists were moved 'to devote themselves to statistical anthropology'.[15] Anthropology and 'eugenics',
'a popular word', were linked in the contemporary controversy over
the idea of 'social progress', which also involved the work of De Vries
and Rignano.[16] In America C. B. Davenport drew on anthropological
material in his popular *Heredity in Relation to Eugenics* which, with
Bergson's *Creative Evolution*, was one of 1913's 'much discussed
books'.[17] Successive periodical articles show a continuing concern
with social hygiene and human potential, both drawing on anthropological material, and, in turn, affected by it.[18] Havelock Ellis's
eugenic theories were being applied to anthropology during Eliot's
Harvard years.[19]

Material about the transmission of human genetic codes was, in
turn, bound up with the study of heredity in other organisms. While
anthropologists might be studying the worship of totemic animals
and plants and the development of such religion into classical and

---

[13] 'Science Lost in the Sciences', unsigned article, *Nation*, 10 Feb. 1910, p. 131.
[14] Quoted in Albert G. Keller, 'William Gordon Sumner', *Nation*, 21 Apr. 1910,
p. 398.
[15] 'Eugenics and Happiness', unsigned article, *Nation*, 25 July 1912, p. 75:
'Science', unsigned article, *Nation*, 26 June 1913, p. 648.
[16] 'Science. Some Books on Evolution', unsigned piece, *Nation*, 5 Dec. 1912,
pp. 543–4.
[17] Raymond Pearl, 'Controlling Man's Evolution', *Dial*, 16 July 1912, pp. 49–51
(the quotation is from p. 49); advertisement, *Nation*, 3 Apr. 1912, p. 345.
[18] See articles in *Dial*, 16 Oct. 1912, pp. 287–9; *Dial*, 1 Oct. 1914, pp. 249–50;
Nation, 18 July 1912, p. 64; *Nation*, 25 July 1912, pp. 75–6; *Nation*, 14 Aug. 1913,
pp. 137–8; *Nation*, 4 Dec. 1913, p. 526.
[19] See unsigned review of *Anthropological Essays Presented to Edward Burnett
Tylor*, *Nation*, 21 May 1908, p. 470.

other forms of worship, so biologists were studying the evolution of the plants themselves. By 1912 Raymond Pearl could write in the *Dial*

The story of the 're-discovery' of Mendel's work by three European investigators at about the same time (1900) is familiar to everyone . . . It is a conservative statement to say that in . . . twelve years a broader comprehension of the meaning of heredity, and a deeper insight into the laws of inheritance, have been gained than from all the previous investigation and speculation about these basic problems.[20]

Eliot knew how important Mendel's work was, and he did not forget about it. In 1916 he was reading about 'the theorists of biological adaptation: Lamarck, Darwin, Weismann, De Vries, Mendel'.[21] In 1918, writing of a poet's education and apparently toying with notions of poetic evolution, Eliot wrote that there was a close analogy between poetry and science. Poets, like scientists, were contributing to the organic development of culture and poets had to know their predecessors' work, just as modern scientists had to know the work of Mendel and De Vries. The poet who remained with Wordsworth was akin to a hypothetical scientist who had ignored all science after Erasmus Darwin.[22] Eliot did not remain with the science of Erasmus Darwin, and at Harvard he could hardly avoid being aware of contemporary developments, some of which lent themselves to sensational 'copy'.

Involved with Mendelian heredity, visited by and written about by De Vries, Luther Burbank was a plant breeder over whom there was considerable controversy around 1910 and beyond. Fêted far beyond his native state of California, which celebrated 'Burbank Day' in his honour, Burbank was a much-discussed figure, seen by some as a pseudo-scientist, but hailed by the *Nation* as 'the most ingenious and successful of all hybridizers'.[23] As creator of such oddities as the 'white blackberry' and the 'plumcot', Burbank was part of a world threatened by a 'Mélange Adultère de Tout'. W. S. Harwood's 1905 account of his *Life and Work* hailed Burbank as 'the foremost plant-breeder in the world'. His name continued to appear on publishers'

---

[20] Raymond Pearl, 'Recent Discussions of Heredity', *Dial*, 16 May 1912, p. 397.
[21] Review of L. M. Bristol, *Social Adaptation*, *New Statesman*, 29 July 1916, p. 405.
[22] 'Contemporanea', *Egoist*, June–July 1918, p. 84.
[23] 'Burbank's Improved Fruits', unsigned article, *Nation*, 13 July 1911, pp. 39–40.

lists until at least 1939, when *An Architect of Nature*, a revised
shorter version of his autobiographical *The Harvest of the Years*,
appeared in 'The Thinker's Library' listed alongside such works as
Tylor's *Anthropology*, Thomson's *City of Dreadful Night*, and
Frazer's *Adonis*. Such a collocation of titles is easily explained if we
realize that all these books could be classed together with others on
the list including some by Haeckel, Charles Bradlaugh, and Havelock
Ellis as hostile to, or at least sceptical about, conventional religion.[24]
Burbank's ideas did not stop at plant life in the conventional sense
but went on to deal with 'improving the human plant'.[25]

Eliot's 'Burbank with a Baedeker' is pointedly about the hybridiza-
tion of the human plant 'Chicago Semite Viennese', and contemporary
science allows him a biologically vast sweep over the gaze which
at once uneasily connects and uneasily separates the aboriginally
primitive from the sophisticatedly civilized painter of the elaborate city,
when crude eye in 'protozoic slime' stares at Canaletto's perspective.
The Davidsonian adjective 'protozoic' points to contemporary science
where plasm was being studied intently when Eliot was a student.
'All the inborn characters of an individual take origin in the germ-plasm
whence he sprang', quotes a 1906 *Nation* reviewer, before going on to
mention Burbank.[26] Henry Smith Williams, in his study of Burbank,
before explaining 'the idea that the racial germ-plasm conveys the
record of past environments and predestines the character', had
presented Haeckel's account of 'the fundamental law of heredity'
according to which the developing human embryo passed through the
stages of fish and lower mammals 'to reproduce the ancestral forms
through which the race has passed in its evolutionary progress'.[27]
In Eliot's early poetry we can find the presence of such a theory in
the masturbatory verse of 'The Death of Saint Narcissus':

> Then he knew that he had been a fish
> With slippery white belly held tight in his own fingers,
> Writhing in his own clutch, his ancient beauty
> Caught fast in the pink tips of his new beauty.

[24] W. S. Harwood, *New Creations in Plant Life: An Authoritative Account of the Life and Work of Luther Burbank* (New York and London: Macmillan, 1905), p. 1.
[25] See *Luther Burbank: His Methods*, etc. (New York: Luther Burbank Press, 1914–15; 12 vols.), xii. 203.
[26] 'Notes', unsigned article, *Nation*, 26 Apr. 1906, p. 345.
[27] H. S. Williams, *Luther Burbank* (London: Grant Richards, 1916), p. 273.

This peculiarly vivid self-regard of the developed form conscious of the undeveloped form of himself echoes Prufrock's familiar outburst,

> I should have been a pair of ragged claws
> Scuttling across the floors of silent seas.

The presentation of a reversion to the primitive evolutionary history of the race promises an escape from the complexities of civilized life.

As a graduate student of philosophy Eliot was well aware of interests in evolutionary theory. In Paris in 1910–11 he had been momentarily converted to Bergsonism and discussed avidly with Aiken *L'Évolution créatrice*.[28] This, Bergson's most recent book, connected the primitive and the sophisticated, linking heredity in plants to heredity in man—all part of one surge of *élan vital*. Bergson's ideas of 'prenatal dispositions' and unconscious memories smuggled through 'the half-open door' would be with Eliot until at least *Four Quartets*.[29] In Bergson's work larger considerations of evolution from the simplest organisms upward could form an introduction to anthropological writing which dealt with the relations between primitive and civilized man. Using Darwin and De Vries, dealing with ideas of adaptation, heredity, and mutation, he discusses a guiding principle of evolution, a continuing *élan vital* which makes evolution a partly conscious process of continuing creation. The work of De Vries seemed to have shortened the time required for evolutionary changes to take place. Bergson's idea of a continuing original impetus passed on through the living cells meant that, for instance, 'no definite characteristic distinguishes the plant from the animal' since each shares some feature of the other. But vegetable, animal, and rational are seen not as developing one from another, but as three *'divergent directions of an activity which has split up as it grew'* (Bergson's italics).[30]

By 1913 Eliot had reacted sharply against Bergson's pseudo-science, remembering satirically how the Frenchman wrote very entertainingly on the structure of a frog's eye.[31] But Bergson's work had a lasting influence and was part of an intellectual climate where humanity was linked continually to the life of lower organisms. Eliot

[28] Eliot cited in Gray, p. 2; Aiken, *Ushant*, p. 157.
[29] Henri-Louis Bergson, *Creative Evolution* (1907) trans. Arthur Mitchell (London: Macmillan, 1911), p. 5.
[30] Ibid., pp. 25, 92, 111, and 142.
[31] 'Eeldrop and Appleplex', *Little Review*, May 1917, p. 10.

recalled the philosophical impact of the work of Walter Pitkin, fashionable in 1912 with his talk of 'the adaptation of the flatfish'.[32] The following year Conrad Aiken wrote to Eliot that 'all philosophy should start with a study of biology, morphology, and such "literature" as The Origin of Species'.[33] Eliot was not entirely heedless of such advice. In 1913–14 he attended a seminar series which included a paper concerned with heredity, discussing Mendel's ideas of variations being predetermined by the make-up of the germ cells which formed traceable links between generations. This paper related Mendel's ideas to human heredity, dealing with the work of Galton, Davenport, and Pearson.[34] At Harvard Eliot learned some physiology.[35] In 1915 at Oxford he attended lectures on 'Mental Evolution' which dealt with primitive man and the biological theories of Darwinian pangenesis, Weismann on germ-plasm, De Vries on mutation, and Mendel on heredity.[36] He would possess a book on *Intracellular Pangenesis*, as well as Rignano's *Upon the Inheritance of Acquired Characters*.[37] In 1918 Eliot would draw attention to the 'exceptional importance' among 'recent periodical literature of ethical interest' of a series of articles by a Professor McBride entitled 'Study of Heredity', along with other eugenic pieces by various writers including Havelock Ellis's 'Birth-Control and Eugenics'. In this review Eliot quotes a passage from one of McBride's articles which points out that

In all cases where large numbers of a given species of animals are raised under somewhat artificial conditions a certain number of monsters will be produced, apparently owing to a disturbance of the germ-cells in their growing and ripening. This is true both of insects raised on banana peel and of human beings raised in a large city.[38]

[32] 'Views and Reviews', *New English Weekly*, 6 June 1935, p. 151.
[33] *Selected Letters of Conrad Aiken*, ed. Joseph Killorin (New Haven: Yale University Press, 1978), p. 30 (letter to T. S. Eliot, 'Rome [March 1913]'.
[34] Grover Smith, ed., *Josiah Royce's Seminar, 1913–1914: As Recorded in the Note-books of Harry T. Costello* (New Brunswick, NJ: Rutgers University Press, 1963), pp. 10, 25, 26.
[35] Three pages of Eliot's notes on the physiology of skin, probably taken during H. S. Langfeld's course on Experimental psychology in October 1911, survive in Houghton.
[36] These notes, dating from 29 Apr. to 19 May 1915 are on pages at the back of Eliot's notes on Aristotle (Houghton). The lecturer was W. McDougall.
[37] From a partial list of Eliot's library typed around 1933. Now among Vivienne Eliot's papers in Bodleian.
[38] 'Recent British Periodical Literature in Ethics', *International Journal of Ethics*, Jan. 1918, p. 273.

Such a passage may have had personal significance for the Eliot who
had complained to Aiken in late 1914 of being very self-conscious
in a large city and of having nervous attacks of a sexual nature
when alone there.[39] But it is certainly important for the disturbed
sexuality of *The Waste Land* and of those monsters Sweeney and
Tiresias. In 1914 Eliot wondered if it might be best to become a
post-office clerk and come to terms with failure.[40] He admired
Conrad Aiken's *Earth Triumphant*, but its familiar city clerk was
trapped in the sort of outmoded 'poetic' world which Eliot loved
to destroy with a shocking, accurate image—an image like that
which he deployed in criticizing a friend's story: 'I think your young
man was not as innocent as you pretend; I think it is highly probable
that he always carried "rubber goods" in his hip pocket.'[41] Eliot
would depict such a world in his seedily clerkly London.

> Mr. Eugenides, the Smyrna merchant
> Unshaven, with a pocket full of currants
> C.i.f. London: documents at sight,
> Asked me in demotic French
> To luncheon at the Cannon Street Hotel
> Followed by a weekend at the Metropole.

Hairy, vulgar-speaking, homosexual Eugenides is the antithesis of
the good breeding his name pronounces. A grotesque critique of
eugenic ideas permeates *The Waste Land* in the horror of

> It's them pills I took, to bring it off, she said.
> (She's had five already, and nearly died of young George.)
> The chemist said it would be all right, but I've never been the same.
> You *are* a proper fool, I said,
> . . .
> What you get married for if you don't want children?

The emotionless lovemaking of clerk and typist should be set against
the blandness of such titles as Havelock Ellis's *The Task of Social
Hygiene*. Though scarcely 'Polyphiloprogenitive', *The Waste Land*
is a poem possessed with breeding and its pain and with the apparent
changes in religion and culture. Both these aspects of the poem link
a panoramic concern to a personal agony. If the story of Mendel's

[39] Letter to Aiken, 31 Dec. 1914 (Huntington).
[40] Ibid., 30 Sept. 1914 (Huntington).
[41] Ibid., 21 Nov. [1914] (Huntington); *Ushant*, pp. 114–15.

researches into breeding and the origins of acquired character stayed with him from his youth, Eliot also maintained an interest in other, perhaps more complex, origins. Recent bibliographical evidence shows that in 1918 he was also writing on a series of booklets by Rendel Harris, published under the general title *The Ascent of Olympus*.[42] Harris's studies of the evolution of classical divinities from plants ranged widely across the ancient and modern worlds moving, as Eliot wryly noted, from the Vedic Soma to ceremonies practised at Lincoln College, Oxford. The tongue-in-cheek quality of Eliot's (anonymous) review gives way to a grin in the concluding paragraph of his review, when he sees Harris's own scholarship in terms of plant-breeding: 'If Dr. Harris is ever found to be wrong, it will be because he clings tenaciously to a single (vegetable) line of descent. This line is certainly traceable, but it seems possible that a developed god may have an extremely complex parentage, not to be reduced to a single root.'[43] Yet *The Waste Land* was to show that Eliot could make very different use of the sort of ideas put forward by Rendel Harris. That poem, before going on to make use of various fertility rituals and references to ancient divinities, begins quite literally with the not unrelated Burbankian theme of plant breeding.

> April is the cruellest month, breeding
> Lilacs out of the dead land, mixing
> Memory and desire, stirring
> Dull roots with spring rain.
> Winter kept us warm, covering
> Earth in forgetful snow, feeding
> A little life with dried tubers.

For a poem about the presence of the past in continual waning and resurgence, this opening is suitably Mendelian. *The Waste Land* goes on to present the uprooted and the rootless and the infertile (as in 'Gerontion'). Its 'Notes' ask readers to see this not in the context of the pain and despair which sound so often through a violently emotional poem, but in the context of 'vegetation ceremonies'.[44]

---

[42] Elizabeth R. Eames and Alan M. Cohn, 'Some Early Reviews by T. S. Eliot (Addenda to Gallup)', in *Papers of the Bibliographical Society of America*, 70 (1976), pp. 420–4.

[43] Review of Rendel Harris, *The Ascent of Olympus*, *Monist*, Oct. 1918, p. 640.

[44] *CPP*, p. 76.

Recently it has been implied that any anthropological framework
of the poem is something grafted on to it at a late stage and so, the
implication goes, of little importance.[45] But the ideas of breeding,
sexuality, and anthropology are connected in *The Waste Land* so
that they interpenetrate one another just as they did in the Harvard
of Eliot's student days. Such varied themes converged when as a
young man Eliot read anthropological works. The world of the
savage would be brought to confront the world of the city in his
early poetry.

At Eliot's Harvard 'random browsing' was rife.[46] His own courses,
though broadly chosen, hardly represent an abandonment of traditional disciplines and scarcely ventured beyond the arts.[47] But Eliot
moved in a group whose intellectual horizons, though sometimes
cloudy, were broad. One Harvard friend with whom he conversed
late into the night included in the dialogue of a 1908 short story
a variety of topics aired among those interested in philosophy at the
time. The themes range from psychical research, Buddhism, and the
transmigration of souls to 'evolutionary theory . . . Huxley and
Haeckel'.[48] Just as some interest in evolutionary theory seems to
have been *de rigueur* in Harvard at the time, so, among Eliot's cricle,
does some interest in the East. Conrad Aiken, with whom Eliot
frequently exchanged poems in his student days, described a Burmese
sculptor who carved an image that became his god. Java was the
fashionable 1890s landscape of a contemporary *Advocate* story.
Eliot was to some extent influenced by his friends, but his gaze was
more penetrating. Aiken wrote poems about city clerks and with titles
such as 'To a City Evening'.[49] The *Advocate*, with Eliot on its
editorial board, published verse on 1890s themes like 'Loafers in the
Park', while Tinckom-Fernandez enthused over Lionel Johnson and
complained in verse to 'The Street Organ' that when 'Spring shyly
taps the window-pane', none the less

---

[45] This I take to be the implication of, A. Walton Litz, '*The Waste Land* Fifty
Years After', in Litz, ed., *Eliot in his Time* (London: Oxford University Press, 1973),
pp. 3–22.
[46] 'Thoroughness in College', unsigned article, *Nation*, 12 Dec. 1912, p. 558.
[47] Most of these are listed in Matthews, pp. 27–8.
[48] *Ushant*, p. 136; F. Schenck, 'Psychical Research', *Harvard Advocate*, 84
(1908), p. 119.
[49] Conrad Aiken, 'The Burmese Sculptor', *Harvard Advocate*, 86 (1908), p. 9;
'The Gentlemen from Java', unsigned story, ibid., p. 109; Conrad Aiken, 'To a City
Evening', *Harvard Advocate*, 90 (1910), pp. 6–7.

> Still in my heart the ancient pain,
> Where, in my heart, old memories lie,
> Touched by the songs an organ plays.[50]

All these factors can be related to Eliot's poetry. But in their diction, they are trapped in a verbose and over 'poetic' Nineties style. Just as Eliot responded to the sharpest passages of Davidson, so he responded to the contemporary interest in the East (an interest quickened in him by Fitzgerald, Kipling, and *The Light of Asia*), but in a way that owed less to Arthur Symons's Javanese dancers than to a viewpoint which should be described as anthropological.

Eliot read widely at Harvard in mysticism and related subjects. A file of sixty annotated index cards from this period contains over ninety separate titles of books and articles.[51] Just over half the cards are devoted to Christian subjects, and the titles include Kellner on Yoga, Koepper on the Buddha, and B. P. Blood on anaesthetic revelation. Eliot took his most copious notes (on eight cards) from F. B. Jevons's *Introduction to the History of Religion* which dealt almost exclusively with primitive religion, though it also touched on 'features common to both civilized man and existing savages, or rather to their ancestors'. Using the chapter headings of Jevons's book, Eliot's themes included 'Taboo', 'Totemism', 'Tree and Plant Worship', 'The Transmigration of Souls', and 'The Mysteries'. Various details caught his eye, such as 'the worship of stones' being a degeneration of the altar.[52] This, blended with a passage of Thomson (see chapter II above), would play its part in the degeneration of 'The Hollow Men' when 'Lips that would kiss / Form prayers to broken stone'. Eliot's various anthropological interests might appear relevant to that poem with its '*Here we go round the prickly pear*'. Frazer and others taught him how rituals degenerated into childishness, and Eliot twice noted the title of an article on 'Ecstasy and Dance Hypnosis among the American Indians', but, for now, such points are important because they show the general emergence of his interest in the primitive. Among other works he

---

[50] H. Nickerson, 'Loafers in the Park', *Harvard Advocate*, 87 (1909), p. 76; W. G. Tinckom-Fernandez, 'Book Review' (*Selected Poems of Lionel Johnson* (London: Elkin Matthews, 1909)), *Harvard Advocate*, p. 132; Tinckom-Fernandez, 'The Street Organ', ibid., 85 (1908), p. 152.

[51] Houghton, partly detailed in Gordon, pp. 141–2.

[52] Frank Byron Jevons, *An Introduction to the History of Religion* (London: Methuen, 1896), pp. 6, vii, 142.

studied were R. R. Marett's *The Threshold of Religion* (1909), Irving King's *The Development of Religion* (1910), and James Woods's *Practice and Science of Religion* (1906). Woods taught Eliot Greek Philosophy in 1911–12, introducing him to Heraclitus.[53] Eliot's card file contains several cards of notes on the early Greek philosophers. Later he would spend 'a year in the mazes of Patanjali's metaphysics under the guidance of James Woods'.[54] From a clerical and academic background similar to Eliot's, Woods, before becoming a professor of philosophy at Harvard, had been interested in comparative religion and psychology and had been both an instructor and a graduate student of anthropology in Eliot's university. Woods's *Practice and Science of Religion* contained chapters on 'Primitive Beliefs', 'Mystical Ideals', and 'Levels of Religion'. In it Eliot came across ideas which would be of crucial importance to his own work. One of these, to which he would return many times, dealt with the development of drama and its relations with primitive ritual. Woods had related dramatic forms to those of primitive religion and had written that 'the god is alive, as alive as Hamlet, alive with the beliefs of his worshippers. So the concept of the god also consists of objectified feelings.'[55] In 1919 Eliot was writing in his own essay, 'Hamlet', of the 'objective correlative' necessary to all art.[56] His literary theory was drawing on his anthropological reading.

For the moment, though, the most important aspect of Woods's book is its title. Eliot was scientifically interested in religion. He was now in revolt against his Unitarian upbringing, an upbringing 'outside the Christian Fold', and one in which things were 'either black or white', which is not quite the same as good or evil.[57] An atmosphere of 'intellectual and puritanical rationalism' had prevailed where 'The Son and the Holy Ghost were not believed in, certainly; but they were entitled to respect.'[58] If a respecting, respectable surface characterized his childhood religion, the same characteristics were found in the tasteful composure of Boston. With its 'Puritan traditions of Beacon Hill' (home of Adeline Moffat),

[53] Eliot's notes on Woods's 1911–12 class are in Houghton.
[54] *ASG*, p. 40.
[55] James Haughton Woods, *Practice and Science of Religion: A Study of Method in Comparative Religion* (New York: Longmans Green, and Co., 1906), p. 57.
[56] *SE*, p. 145.
[57] 'Books of the Quarter', *Criterion*, July 1931, p. 771.
[58] *A Sermon* (Cambridge: The University Press for Magdalene College, 1948), p. 5; 'Books of the Quarter' (C322), p. 771.

Boston 'was and is quite uncivilized but refined beyond the point of civilization'.[59]

Eliot's reaction was to set against this world a fascination with hidden depths of emotion and the revelation of such emotion in ways far alien to, and more potent than, the Bostonian world around him. He wished to move beyond gentility and genteel religion, both of which he analysed bitingly; at least philosophy other than that 'taught by retired non-conformist ministers' had the virtue of being 'on the whole anti-religious, which was refreshing'.[60] Refreshed, Eliot became not antireligious, but stayed at once both scientifically distant from and passionately fascinated with religion's visionary violence. The practice and science of religion which interested him took in far more than the Boston Unitarian or even the Christian Church. In a chapter on 'Primitive Beliefs' Woods had written of 'The brilliant work in Ceylon of two German naturalists, the brothers Sarasin [which] gives us a description of one of the simplest attempts to express in an action a permanent object of desire.' The Rock-Veddahs, whose customs were described by the Sarasin brothers, were called by Woods 'one of the five or six most primitive peoples alive to-day' and one of the Veddah's ceremonies was described by him in some detail.

In the crises of his life, when there is sickness or childbirth or preparation for the hunt, there is worship of the arrow. In the forest at night the men dance in a circle around the arrow stuck tip down in the ground. There are fires to add to the festive character of the scene. There are rhythmic cries, expressions of thanks or of hope. This continues until all are worked up to almost convulsive excitement. But there is no clear concept of an arrow-spirit, or of a god, or of any soul. The dance expresses what they want; but the ideal is not so clear that it can be expressed in a concept or in a word or in a figure.[61]

Eliot developed an interest in the origins of such rites and in 1913 wrote a paper on the interpretation of primitive ritual which was to have lasting importance for his work.[62] He read widely in anthropological

[59] 'American Critics', *TLS*, 10 Jan. 1929, p. 24; 'The Hawthorne Aspect', *Little Review*, Aug. 1918 (rpt. Edmund Wilson, ed., *The Shock of Recognition* (London: W. H. Allen, 1956), p. 860).

[60] 'Views and Reviews' (C387), p. 151.

[61] Woods, n. 55 above, p. 47.

[62] This paper was rediscovered by Piers Gray in the Hayward Bequest, King's College, Cambridge.

scholarship, and at least since his time at Harvard was intensely interested in convulsive and hallucinatory violence of the most savage sort. This he imported into an urban setting. So Saint Narcissus in an early poem

                              became a dancer before God
    If he walked in city streets
    He seemed to tread on faces, convulsive thighs and knees.

But Eliot was to be haunted by this dance of the Veddahs. He read about it again in Wundt's *Elements of Folk Psychology* which he reviewed in 1917. In the chapter on 'Primitive Man', Wundt wrote of magical ideas associated with external objects.

The magical significance has, of course, frequently disappeared from the memory of the natives. The Sarasins saw the Veddahs execute dances about an arrow that had been set upright. On enquiring the reason, they were told: 'This was done even by our fathers and grandfathers; why should we not also do it?' . . . Those ceremonies particularly that are in any way complicated are passed down from generation to generation, being scrupulously guarded and occasionally augmented by additional magical elements . . . The meaning of the ceremonies has for the most part long been lost to the participants themselves, and was probably unknown even to the ancestors . . . The conditions here are really not essentially different from those that still prevail everywhere in the cult ceremonies of civilized peoples. It is the very fact that the motives are forgotten that leads to the enormous complexity of the phenomena.[63]

In 1928, Eliot, pondering 'the relation of drama to religion', wrote of the work of Harrison, Cornford, and Gilbert Murray, as well as 'the antics of the Todas and the Veddahs'.[64] Eliot was no anthropologist, and his memory for names was fallible. But anthropology seeded his imagination, giving him a tactical weapon in his literary struggles. Eliot was haunted by this Veddah ceremony, pointing out in 1926 that dramatic form could occur at various stages along a line stretching from liturgy at the one end to realism at the other. He saw 'the arrow-dance of the Todas [sc. Veddahs]' and the work of Sir Arthur Pinero as being at the opposite extremities

[63] Wilhelm Wundt, *Elements of Folk Psychology*, trans. Schaub (London: Allen & Unwin, 1916), pp. 90–1.
[64] *SE*, p. 44.

of such an imagined line.[65] Here Eliot is looking back towards
Wundt, his seminar papers, and also towards *The Waste Land* where
the poet who said, when writing of anthropology, primitive poetry,
and savagery in 1919, 'The maxim, Return to the sources, is a good
one', is trying to do something to move towards a more primitive
conception of poetry and drama.[66] Eliot had used the same phrase
when praising Charles Péguy's stand for 'a real re-creation, a return
to the sources' in 1916.[67] As it had been in his childhood days of
Mayne Reid, Eliot's attitude towards primitive man was ambivalent.
The savage could seem intensely foolish, devoting great energies to
inane projects. At other times he could seem a vital origin, a source
to be seriously examined and tapped. In 1923, writing of 'savage
or primitive art', in a wider context, Eliot had stressed that 'literature
cannot be understood without going to the sources: sources which
are often remote, difficult, and unintelligible unless one transcends
the prejudices of ordinary literary taste.[68]

'Taste' was something which Eliot was concerned with sending
up. In 'Portrait of a Lady', a suppressed violence lies beneath the
'life composed so much, so much of odds and ends', in the tomblike
atmosphere where life is squeezed out of the stalks of lilac in a modern
parody of a vegetation ceremony. As a child, Eliot had been excited
by Dowson's 'Impenitentia Ultima' where 'the viols in her voice be
the last sound in mine ear', but now Eliot sharply undermines such
1890s sentiments when in 'Portrait of a Lady', 'The voice returns
like the insistent out-of-tune / Of a broken violin on an August
afternoon.'[69] Against the phoney culture of the Bostonian world,
Eliot found something both intimate and savage.

> Among the windings of the violins
> And the ariettes
> Of cracked cornets
> Inside my brain a dull tom-tom begins
> Absurdly hammering a prelude of its own,
> Capricious monotone
> That is at least one definite 'false note.'

[65] 'Introduction' to Charlotte Eliot, *Savonarola: A Dramatic Poem* (London:
R. Cobden Sanderson, (1926)), p. x.
[66] 'War-paint and Feathers', *Athenaeum*, 17 Oct. 1919, p. 1036.
[67] 'Charles Péguy', *New Statesman*, 7 Oct. 1916, p. 20.
[68] 'The Beating of a Drum', *Nation & Athenaeum*, 6 Oct. 1923, p. 11.
[69] 'On Teaching' (C640), p. 78; Dowson, *Poetical Works*, p. 81.

The thought that Eliot, the Donne-lover, should play on his own name in his own poem is one which it would be wrong to exclude. In Eliot's draft the tom-tom had been firstly 'droll', then 'strong' and 'male'.[70] Eliot's final choice of 'dull' suggests not so much ordinariness (tom-toms are uncommon in drawing-rooms) as powerful bluntness. Certainly the tom-tom's comic ('Absurdly') yet deep ('Inside my brain') and ferocious ('hammering') sound provides the needed savage counter-prelude to be set against the Chopin which the lady has taken over. The tom-tom is a crude image from the sort of crude adventure world of 'the comics or the sporting page' (or possibly Harvard's Peabody Museum). Salvation is to be sought in a return to the jungle. It is the tom-tom that convinces the reader that the young man is, potentially at least, capable of being saved from the genteel hell in which he is immured. But he is not saved at the poem's end, because he cannot achieve the smile which would be a savage smile because it would be a smile of triumph breaking the power of a dead woman. Eliot's narrator cannot win free because he is like Eliot's Henry Adams who 'had been too respectful of whatever was important, he laughed at nothing'.[71] Prufrock is killed by the 'human voices' because his savagery is inarticulate in the extreme, a matter only of wriggling, spitting, and biting. He can find no language in which to articulate the inner energy whose potential is at times sensed in his words. But words are in the control of his female enemies, who all the time may say,

> 'That is not it at all,
> That is not what I meant, at all.'

Marvellously eloquent in a language which stupefies and defeats him, Prufrock's release lies in bursting out of words in that primitive shape—

> I should have been a pair of ragged claws
> Scuttling across the floors of silent seas.

But that is an escape route which he is unable to pursue. Eliot watches with scientific yet painful fascination the Prufrock who is in part himself. The tom-tom hints that the young man of 'Portrait of a Lady' is a little more powerful than Prufrock, but he is still not strong

[70] The Complete Poems MS (Berg).
[71] 'A Sceptical Patrician', *Athenaeum*, 23 May 1919, p. 362.

enough to smile the final, propriety-breaking smile. Laughter here is one of Eliot's most important weapons. His first book is a poetry of small outbursts, looking towards hope for the future when, in the destruction of an outworn way of life, 'the spirit unappeased and peregrine' shall win through

> And the fullfed beast shall kick the empty pail.
> For last year's words belong to last year's language
> And next year's words await another voice.

The most promising outbursts are funny ones, such as those against fake life and fake talk in 'Portrait of a Lady' and against the dead life of 'Aunt Helen':

> The Dresden clock continued ticking on the mantlepiece,
> And the footman sat upon the dining-table
> Holding the second housemaid on his knees—
> Who had always been so careful while her mistress lived.

Elsewhere, outbursts are suppressed because too weak,

> When evening quickens faintly in the street,
> Wakening the appetites of life in some
> And to others bringing the *Boston Evening Transcript*.

Other break-outs come in the shape of a too crude violence which tears smiles from faces ('Morning at the Window') or results in that melodramatic last line of 'Rhapsody on a Windy Night'. Only the brilliant barbarian, Mr Apollinax, who is associated with the primitive impropriety of 'Priapus' and whose depths are revealed in the fact that he 'laughed like an irresponsible foetus. / His laughter was submarine and profound' finds the way out of a stultifying world where 'Channing' is ironically coupled with 'Cheetah'. He has on his side the sympathy and envy of the head-hunting poet who

> looked for the head of Mr. Apollinax rolling under a chair
> Or grinning over a screen
> With seaweed in its hair.

The poet himself, however, has still to win his freedom, because though the tools of poetic analysis will aid him, he has not yet found enough of the right words to talk his way out.

If elements of primitivism were entering Eliot's poetry, it should not be assumed that he was simply a would-be savage. He was sensitive about his interests in the primitive. He *was* concerned about the avant-garde primitive, admiring modern sculpture in 1915 by likening it to the vigorousness of Central African images, and bringing back from Paris, four years earlier, a print of Gauguin's *Crucifixion*.[72] But Conrad Aiken remembered that 'The suggestion that [the Gauguin] . . . was a kind of sophisticated primitivism brought the reply, with a waspishness that was characteristic, that there "was nothing primitive about it." '[73]

Despite such a denial, Eliot's interest in the primitive was growing. His reading on Christian mysticism shows him to have been interested particularly in the physical and often sexual violence associated with extreme religious emotion. He was interested in Murisier's connections between illness and vision, reading Janet on hallucination and hysteria, and was investigating religio-sexual frenzy. He transcribed on to one of his index cards the words of Havelock Ellis in *The Psychology of Sex*, 'Love and religion are the two most volcanic emotions to which the human organism is liable.' Such concerns emerged in poems like 'The Death of Saint Narcissus'. Even expressed in late romantic terms, Eliot had a hunger after and a horrified fascination with 'the voice of men in pain' and flowers whose 'petals are fanged and red / With hideous streak and stain'. Potent, bloody sexuality could furnish an escape from impotent pallor:

> Whiter the flowers, Love, you hold,
>   Than the white mist on the sea;
> Have you no brighter tropic flowers
>   With scarlet life, for me?

Sex and religion break out with a developing savagery that is like a shock treatment which Eliot wishes to administer to himself and the quiet, respectable society of which he is a part:

> So he became a dancer to God.
> Because his flesh was in love with the burning arrows
> He danced on the hot sand
> Until the arrows came.
> As he embraced them his white skin surrendered itself to
>   the redness of blood, and satisfied him.
> Now he is green, dry and stained
> With the shadow in his mouth.

[72] Letter to Isabella Stewart Gardner, 4 Apr. 1915 (Xerox in Houghton); Aiken, 'King Bolo and Others', p. 21.
[73] Aiken, as in preceding note.

Eliot wrote this in 1914/15, once his anthropological reading was
well advanced. There is perhaps a childish violence about it; it
sometimes reads like pulling the wings off flies. But the purpose is
clear. The poet is trying to burn off the page the polite saintliness
of an exhausted way of life, replacing it with a savage rite.

Eliot was fascinated by the idea that such intense experience might
even now be available to modern man. He transcribed from James's
Varieties of Religious Experience the passage which states James's
view that 'our normal waking consciousness, rational consciousness
as we call it, is but one special type of consciousness, whilst all about
it, parted from it by the filmiest of screens, there lie potential forms
of consciousness entirely different'.[74] Eliot noted how ether might
stimulate mystical consciousness, and read James's quotations from
B. P. Blood of visions experienced under ether 'in a bed pushed up
against a window, a common city window in a common city street'.
Eliot also noted the reference to the work of R. M. Brucke, one of
whose accounts of visionary states James quoted:

I had spent the evening in a great city, with two friends, reading and
discussing poetry and philosophy . . . All at once, without warning of any
kind, I found myself wrapped in a flame-colored cloud. For an instant
I thought of fire, an immense conflagration somewhere close by in that great
city; the next, I knew that the fire was within myself.[75]

Such visions could clearly connect with the dark sometimes violent
visions of Thomson and Davidson. The ether at the beginning of
Prufrock introduces us to a series of seedy streets where the speaker
dreams of some ultimate illumination which would let him

> say: 'I am Lazarus, come from the dead,
> Come back to tell you all, I shall tell you all' —

but Prufrock, who does not think the mermaids will sing to him,
hardly seems cut out for this sort of thing. Such visions were frightening, yet fascinating. Both aspects are present in the unpublished
poem, 'Silences', when a city street, filled with talkative waves of
people, divides to afford an ultimate moment when life may be
justified.[76] Eliot, of all poets, was crucially affected by his reading,

[74] Transcribed on index card in Houghton; William James, *The Varieties of
Religious Experience* (1902; rpt., London: Collins, 1960), p. 374.
[75] Ibid., pp. 379 and 385.
[76] The Complete Poems MS dated June 1910 (Berg).

and in the early unpublished poems he continually returns to moments
of revelation, or the lack of it, in squalid urban life. Such moments
are sometimes silent, sometimes they beat like drums on the skull.
They are, in the terms of an early unpublished poem, insights rushing
over desert plains like railway engines.[77] But, just as it would be
wrong to see Eliot simply as Prufrock, so it is wrong to assume
that he himself saw visions in the streets of Boston. Involved in
labyrinthine metaphysics and, later, in the Bradleian concern with
degrees of reality, Eliot may have wished for such visions, but it is
most likely that they came to him largely through the books he read.
  Certainly they were present in the French poets who became so
important to him. A poem such as Baudelaire's 'A une Madone'
treated religio-sexual frenzy in shocking verse,

> Enfin, pour compléter ton rôle de Marie,
> Et pour mêler l'amour avec la barbarie,
> Volupté noire! des sept Péchés capitaux,
> Bourreau plein de remords, je ferai sept Couteaux
> Bien affilés, et comme un jongleur insensible,
> Prenant le plus profond de ton amour pour cible,
> Je les planterai tous dans ton Cœur pantelant,
> Dans ton Cœur sanglotant, dans ton Cœur ruisselant![78]

Eliot's early unpublished poem whose first section was about an
idolatrous lover flogging himself to death and whose second section
has him strangling his mistress in the manner of 'Porphyria's Lover'
would lead to the religio-sexual frenzy of 'The Death of Saint
Narcissus'.[79] As Symons informed him, the Symbolists were pre-
eminently concerned with a powerful world at odds with or beyond
the world of ordinary life. Eliot found how his most disparate
concerns could be united in verse. Laforgue commanded his own
nerves 'Redevenez plasma!' and exhorted

> Faisant de leurs cités une unique Ninive,
> Mener ces chers bourgeois, fouettés d'alléluias,
> Au Saint-Sépulcre maternel du Nirvâna![80]

[77] 'Fragment, Bacchus & Ariadne, 2nd Debate between the Body & Soul', dated
Feb. 1911 in Miscellaneous Material MS (Berg).
[78] Baudelaire, *Oeuvres complètes*, i, ed. Claude Pichois (Paris: Gallimard, 1975),
p. 59.
[79] Miscellaneous Material MS (Berg); the poem is described in Gordon, pp. 27–8.
[80] Jules Laforgue, *Poems*, ed. J. A. Hiddleston (Oxford: Basil Blackwell, 1975),
pp. 63 and 64 ('Préludes Autobiographiques').

Urban, urbane dandy, Laforgue had his characters exclaim

> O femme, mammifère à chignon, ô fétiche,
> On t'absout; c'est un Dieu qui par tes yeux nous triche.[81]

As a modern city man, he none the less had his Parisian sage recommend a course whose images would find their way into Eliot's verse, as well as indicating the direction which Eliot's studies would take as regards Sanskrit, Pali, and Japanese Buddhism.

> Ah! démaillotte-toi, mon enfant, de ces langes
> D'Occident! va faire une pleine eau dans le Gange.
>
> La logique, la morale, c'est vite dit;
> Mais! gisements d'instincts, virtuels paradis,
>
> Nuit de hérédités et limbes des latences!
> Actif? passif? ô pelouses des Défaillances,
>
> Tamis de pores! Et les bas-fonds sous-marins,
> Infini sans foyer, forêt vierge à tous crins![82]

The father of the Symbolists, Baudelaire brought home to Eliot the message of the 'Unreal City'. It is significant that when Eliot quotes the words 'Fourmillante Cité', he gives Baudelaire's 'cité' a capital C, as if it were the 'City' of *The Waste Land*.[83] Baudelaire's was a poetry where religion and sexuality mixed as they did in Eliot's study of mysticism and anthropology. With Baudelaire's celebrations of his black mistress, Jeanne Duval, almost ridiculously powerful sexuality and primitive imagery blend.

> Bizarre déité, brune comme les nuits,
> Au parfum mélangé de musc et de havane,
> Œuvre de quelque obi, le Faust de la savane,
> Sorcière au flanc d'ébène, enfant des noirs minuits . . .[84]

Baudelaire and the Symbolists were at once poets of the urban world and of the savage. For Laforgue, there was 'le rage de vouloir se connaître—de plonger sous la culture consciente vers "L'Afrique intérieure" de notre inconscient domaine'.[85] In Rimbaud's *Illuminations* over which Eliot enthused in 1917, praising provocatively

---

[81] Laforgue, *Poems*, p. 71 ('Complainte des voix sous le figuier boudhique').
[82] Ibid., p. 134 (Complainte du sage de Paris').   [83] CC, p. 127.
[84] Baudelaire, n. 78 above, p. 28.
[85] Jules Laforgue, *Derniers Vers*, ed., M. Collie and J. M. L'Heureux (Toronto: University of Toronto Press, 1965), p. 7; cited in Gray, p. 9.

the combination of a strange precision with an admirable cogency in choice and juxtaposition of images, the city again confronts the savage, blending with it when, in 'Villes', 'Les Bacchantes des banlieues sanglotent' and 'Les sauvages dansent sans cesse la fête de la nuit'.[86] In another of the *Illuminations*, a 'parade sauvage' of Hottentots, Molochs, and others perform plays cleverer than any imagined in 'l'histoire ou les religions'. The savage in Rimbaud's best-known poem signals a release from quotidian life into another intensity.

> Comme je descendais des Fleuves impassibles,
> Je ne me sentis plus guidé par les haleurs:
> Des Peaux-Rouges criards les avaient pris pour cibles,
> Les ayant cloués nus aux poteaux de couleurs.[87]

The Eliot who absorbed Baudelaire, Laforgue, and Rimbaud was the Eliot who wrote of slummy streets and went on in 1913 to study the interpretation of primitive ritual.

Eliot also made the connection between the savage and the city more directly and outrageously in a series of deliberately scandalous verses he wrote (for a trusted circle) about King Bolo and his Big Black Queen. Bolo was another 'man who was king'. His name, as Eliot knew from fine art classes, meant 'ground for gilding'.[88] On the one hand he was the rakish man-about-town, gilded with the sophistication of top hat, monocle, and cigar; on the other, he was lusty ruler with his Big Black Queen over a tribe of 'some primitive people called the Bolovians', who (as Eliot later elaborated to Bonamy Dobrée) 'wore bowler hats, and had square wheels to their chariots'.[89] From one angle, Eliot's couple seem to parody the dandy Baudelaire and his immensely sexual black mistress. Bolo's Queen has a comically gargantuan sexual appetite. Following Baudelaire and the Symbolists, Eliot seems around his graduate student days at least to have seen the negro world as one of greater intensity than the one he knew. So negroes figure in some of his most

[86] 'The Borderline of Prose', *New Statesman*, 19 May 1917, p. 158; Rimbaud, *Œuvres complètes*, ed. Antoine Adam (Paris: Gallimard, 1972), p. 136.
[87] Ibid., p. 126 ('Parade') and p. 66 ('Le Bateau Ivre').
[88] '[Notes on] Fine arts 20b', 22 Feb.–26 May 1910, MS, p. 2 (Houghton).
[89] Eliot drew Bolo in a letter to Aiken 19 July 1914 (Huntington); cp. his drawing of Bolo and his Queen in a letter to Bonamy Dobrée, 12 Nov. 1927 (Brotherton); Bonamy Dobrée, 'T. S. Eliot: A Personal Reminiscence', in Tate, p. 73.

surreal Rimbaud-like prose poems.[90] Another poem's narrator seems envious of a little black girl carrying a geranium. Her certainty of God is beyond his.[91] Finally, in a third poem with a French refrain a black moth ecstatically immolating itself on a candle seems all the more intensely fascinating because it smells of the tropics and comes from Nicobar or Mozambique.[92] Later, in 'Mélange Adultère de Tout', Eliot mocked such naïve exoticism, but Bolo's Big Black Queen draws on it for her potency, a potency of the most carnal sort. She belongs to the Eliot landscape of 'Bullshit' and 'Ballad for Big Louise' to which Wyndham Lewis reacted in 1915: 'They are excellent bits of scholarly ribaldry. I am trying to print them in *Blast*; but stick to my naif determination to have no "words ending in -Uck, -Unt, and -Ugger." '[93] 'You can forward the "Bolo" to Joyce,' Pound cautioned Eliot in 1922, 'if you think it won't unhinge his somewhat sabbatarian mind.'[94] The polite Boston of Adeline Moffat was undermined with straight-faced humour in poems such as 'Aunt Helen'. In the Bolo poems Eliot blew it sky-high. Though he did not add to the poems after 1916, he delighted in elaborating to selected correspondents upon the habits of the Bolovians, often repeating older rhymes.[95] The poems are dirty jokes, satires on 'tasteful' society, caricatures of a Baudelairean world, and, as Bonamy Dobrée (who received many Bolovian communications) realized, partly in their accounts of absurd ritual a pastiche of anthropology. Bolo accompanied Eliot all his life. He is part of the Eliot who loved practical jokes and wrote fan-mail to Groucho Marx.[96] Deliberately outrageous and taboo-breaking, they are a scandalously comic counterpoint to Eliot's serious concerns with religion and sexuality. The absurd version of his reading about religio-sexual frenzy and the evolution of modern rites from primitive rituals is the Big Black Queen who turns up in London as

[90] Miscellaneous Material MS (Berg) (untitled, undated prose poem).

[91] Complete Poems MS (Berg), 'Easter: Sensations of April' (Apr. 1910).

[92] Miscellaneous Material MS (Berg), 'The Burnt Dancer' (June 1914).

[93] *The Letters of Wyndham Lewis*, ed. W. K. Rose (London: Methuen, 1963), pp. 66–7; Lewis to Pound, London '[January 1915?]'.

[94] Ezra Pound, *Selected Letters 1907–1941*, ed. D. D. Paige (London: Faber, 1950), p. 171; Pound to Eliot, 'Paris, [?January]' [1922].

[95] Valerie Eliot, 'T. S. Eliot' (letter), *TLS*, 17 Feb. 1984, p. 165; see letters to Dobrée in Brotherton; Eliot is said to have sent Bolovian material to Theodore Spencer; he enjoyed recalling Bolo 'in conversation and correspondence' with Aiken 'over the years' (Valerie Eliot, ibid.), and with old Harvard friends such as L. M. Little (Houghton).

[96] See, e.g., Gordon, p. 22.

> That airy fairy hairy 'un,
> Who led the dance on Golders Green
> With Cardinal Bessarion.[97]

Mixing the religious and the sexual, the savage and the city were united in this bawdy private epic in ways that prefigured and accompanied the juxtaposition that would play an increasingly important part in Eliot's public poetry.

In 1913 Eliot was asked to present a paper on scientific methodology. His choice of theme, the interpretation of primitive ritual, is a significant one. The paper concentrated on the difference between an objective fact and a subjective interpretation. Both the philosophical and the anthropological materials of the paper were to be of importance to him in the rest of his career. In 1923, while he was completing *Sweeney Agonistes*, he opened a review article called 'The Beating of a Drum' by complaining how little impact Darwin's theories had made on literary criticism. Eliot went on to hint at how anthropology was part of the non-literary material which mattered most to him.

If literary critics, instead of perpetually perusing the writings of other critics, would study the content and criticize the methods of such books as 'The Origin of Species' itself, and 'Ancient Law,' and 'Primitive Culture,' they might learn the difference between a history and a chronicle, and the difference between an interpretation and a fact.[98]

These remarks hark back both to the precise interests of Eliot's paper and to the general interests of its period. When the author of *Primitive Culture* died in 1910, his obituary coupling his name with that of the writer of *Ancient Law* in the *Nation* spoke of how among anthropologists 'Tylor and Sir Henry Maine and the others were, in a sort, pioneers opening to us a fascinating new country.'[99] Maine's book might be termed legal anthropology, and one of his 'phrases' which the *Nation* said had 'become mere commonplaces' was that of the 'legal fiction', which 'conceals, or affects to conceal, the fact that a rule of law has undergone alteration its letter remaining unchanged'.[100] Such an idea is clearly important for a poet such as

[97] Quoted in Aiken, 'King Bolo and Others', p. 22.
[98] 'The Beating of a Drum' (C146), p. 11.
[99] 'Science Lost in the Sciences', unsigned article, *Nation*, 10 Feb. 1910, p. 132.
[100] Unsigned review of Sir Henry Sumner Maine, *Ancient Law*, *Nation*, 14 Feb. 1907, p. 159. Maine, *Ancient Law* (1861), introduced by F. Pollock (London: John Murray, 1906), p. 30.

86 *Savage*

Eliot, continually repeating older forms of words, yet altering their earlier meaning by the way in which he repeats them. In the 1913 paper he examined the same question with regard to primitive rituals. The paper is a complex one, but his own recollection and interpretation of it in 1926 reveals the theoretical aspect which had most lasting impact.

Some years ago, in a paper on *The Interpretation of Primitive Ritual*, I made an humble attempt to show that in many cases *no* interpretation of a rite could explain its origin. For the meaning of the series of acts is to the performers themselves an interpretation; the same ritual remaining practically unchanged may assume different meanings for different generations of performers; and the rite may even have originated before 'meaning' meant anything at all.[101] (Eliot's italics)

In his paper Eliot concentrated on primitive ritual, but the material which he was studying made him well aware that what was true of an aboriginal rain ceremony could also be true of more familiar rites. He knew that Durkheim, for instance, in *The Rules of Sociological Method* points out how

The religious dogmas of Christianity have not changed for centuries, but the role they play in our modern societies is no longer the same as in the Middle Ages. Thus words serve to express new ideas without their contexture changing. Moreover, it is a proposition true in sociology as in biology, that the organ is independent of its function, i.e. while staying the same it can serve different ends. Thus the causes which give rise to its existence are independent of the ends it serves.[102]

Ideas such as this were in turn taken up by anthropologically minded investigators of the development of religion, like Jane Harrison whose *Themis* specifically acknowledges a debt to Durkheim's work.[103] So she stresses in *Themis* which Eliot read for his seminar paper, that

In the study of Greek religion it is all important that the clear distinction should be realized between the comparatively permanent element of the ritual and the shifting manifold character of the myth. In the case before us we have a uniform ritual, the elements of which we have disentangled—the

[101] 'Introduction' to *Savonarola* (B4), p. viii.
[102] Emile Durkheim, *The Rules of Sociological Method* (1895), etc., ed. Steven Lukes, trans. W. D. Halls (London: Macmillan, 1982), p. 121.
[103] Jane Harrison, *Themis* (Cambridge: Cambridge University Press, 1912), p. ix.

armed dance over the child, the mimic death and rebirth; but the myth shifts; it is told variously of Zagreus, Dionysos, Zeus, and there is every variety of detail as to how the child is mimetically killed and how the resurrection is effected. To understand the religious intent of the whole complex it is all important to seize on the permanent ritual factors.[104]

Harrison moved from Greek rites to African tribal dances and back, stressing the essential element of the *dromenon*, an act which is characterized as 'a thing *re*-done or *pre*-done, a thing enacted or represented. It is sometimes *re*-done, commemorative, sometimes *pre*-done, anticipatory, and both elements seem to go to its religiousness.' It is not a simple action, such as hunting or fighting, but a mimesis of that action, 'a desire to *re*-live, to *re*-present' (all italics Harrison's).[105] It is not hard to see the relevance of this to the repeated situations of the quatrain poems.

In 1913 Eliot was in considerable agreement with Harrison's approach. After discussing the various interpretations of a rite, he concluded that

The only part of the fact which can be handled scientifically historically is the ritual. Of this archaeology of the fact Miss Harrison is one of our most proficient exponents. We *can* come to conclusions as to what men did at one period and another, and can to some extent see the development of one form out of another.[106]

Again, though, Eliot was reluctant to commit himself on how far a modern religious form might have developed from a primitive one. That 'to some extent' is the philosophy student's cautious evasion.

Until Piers Gray's rediscovery of this paper, part of which he published in 1982, Eliot's piece was presumed lost.[107] It is now possible to trace in it some of Eliot's anthropological reading. The names of Durkheim and Lévy-Bruhl had been much discussed during 1910–11 when Eliot was in Paris and the primitivism of Matisse and Picasso was also a talking point.[108] Sometimes in his paper Eliot provides only the reference of an author's surname. I have marked these cases with an asterisk, while suggesting the most likely work referred to.

---

[104] Ibid., p. 16.   [105] Ibid., p. 43.   [106] IPR, p. 10.
[107] See Gray, pp. 108–41.   [108] 'A Commentary', *Criterion*, Apr. 1934, p. 452.

88 *Savage*

1913 SEMINAR PAPER REFERENCES

Arthur Bernard Cook.* Eliot mentions Cook as a friend of Jane Harrison who acknowledges his help in the 'Introduction' to *Themis* (see below), pp. xix–xx, where she says his forthcoming book will 'mark an epoch in the study of Greek Religion'. Probably Eliot did not know Cook's work at first hand. His *Zeus: A Study in Ancient Religion* was not published until 1914.

Francis Macdonald Cornford.* 'Chapter on the Origin of the Olympic Games' in *Themis*; Eliot probably knew also his *From Religion to Philosophy: A Study in the Origins of Western Speculation* (London: Edward Arnold, 1912).

Emile Durkheim. *Les règles de la méthode sociologique* (Paris: F. Alcan, Bibliothèque de philosophie contemporaine, 1895).

—— 'Représentations individuelles et représentations collectives', *Revue de métaphysique et de morale*, 1898, pp. 272–302.

J. G. Frazer. *The Golden Bough: A Study in Magic and Religion, Third Edition, Part Three, The Dying God* (London: Macmillan; New York: St Martin's Press, 1911).

Jane Harrison. *Themis: A Study of the Social Origins of Greek Religion* (Cambridge: Cambridge University Press, 1912).

Irving King. *The Development of Religion: A Study in Anthropology and Social Psychology* (New York: Macmillan, 1910). This book makes much use of Spencer and Gillen.

Andrew Lang. *The Making of Religion* (London: Longmans, Green, & Co., 1898).

Lucien Lévy-Bruhl. *Les Fonctions mentales dans les sociétés inférieures* (Paris: F. Alcan, Bibliothèque de philosophie contemporaine, 1910).

R. R. Marett.* 'Is Taboo a Negative Magic?', in W. H. R. Rivers, R. R. Marett, N. W. Thomas, eds., *Anthropological Essays Presented to Edward Burnett Tylor* (Oxford: Clarendon Press, 1907), pp. 219–34.

F. Max Müller. *Natural Religion: The Gifford Lectures Delivered Before The University of Glasgow in 1888* (London: Longmans, Green, & Co., 1889).

Gilbert Murray.* (probably at least his) 'Excursus on the Ritual Forms Preserved in Greek Tragedy', in *Themis*.

E. B. Tylor. *Primitive Culture: Researches into the Development of Mythology, Philosophy, Religion, Language, Art, and Custom*, 2 vols., Third Edition (London: John Murray, 1891). As Gray points out, the pagination of this edition accords with Eliot's references.

It may be of use to set beside these works the following titles which Eliot would review before the publication of his quatrain poems, since they demonstrate his continuing interest in anthropological scholarship. I have omitted various books on eastern religions, and instance only those books which deal with 'savages'. As shall be shown later, it is probable that Eliot's reading went beyond the titles mentioned.

1914–1918

1914–1916. During this period Eliot read Durkheim's *Les Formes élémentaires de la vie réligieuse: le système totémique en Australie* (Paris: F. Alcan, Bibliothèque de philosophie contemporaine, 1912).

July 1916. Review of L. M. Bristol, *Social Adaptation: A Study in the Development of the Doctrine of Adaptation as a Theory of Social Progress* (Cambridge, Mass.: Harvard University Press; London: Humphrey Milford, 1915), *New Statesman*, 29 July 1916, p. 405. This book deals with Darwin, Mendel, Durkheim, and others. On the same page also appears a review of the Webb book detailed below. Gray contends correctly that Eliot is author of this review.

August 1916. 'Durkheim', review of *Les Formes élémentaires* in English translation, *The Elementary Forms of the Religious Life*, tr. Joseph Ward Swain (London: Allen & Unwin, 1915), *Saturday Westminster Gazette*, 19 August 1916, p. 14.

October 1916. Review of Clement C. J. Webb, *Group Theories of Religion and the Individual* (London: Allen & Unwin, 1916), *International Journal of Ethics*, 27, 1, pp. 115–17. Webb summarizes and attacks work of Durkheim, Lévy-Bruhl, Cornford, and Harrison.

January 1917. Review of Wilhelm Wundt's *Elements of Folk Psychology: Outlines of a Psychological History of the Development of Mankind* (tr. Schaub, London: Allen & Unwin; New York: Macmillan, 1916), *IJE* 27, 2, pp. 252–4.

July 1917. Review of Stanley A. Cook, *The Study of Religions* (London: A. & C. Black, 1914), *Monist*, 27, p. 480.

January 1918. 'Recent British Periodical Literature in Ethics', review of various pieces including E. W. McBride, 'The Study of Heredity' (appeared in *Eugenics Review* in four parts between April 1916 and January 1917), *IJE* 28, 2, pp. 270–7.

Another, substantially different, review of Durkheim's *Elementary Forms*, English translation (see above), *Monist*, 28, pp. 158–9.

Revision of above review of Wundt, *Monist*, 28, pp. 159–60.

October 1918. Review of Rendel Harris, *The Ascent of Olympus* (Manchester: Manchester University Press; New York and London: Longmans, Green, & Co., 1917), *Monist*, 28, p. 640.[109]

---

[109] Bibliographical information from four sources: Donald Gallup, *T. S. Eliot: A Bibliography* (London: Faber & Faber, 1969); Eames and Cohn (see n. 42 above);

Much of the material in these reviews grew from the seminal paper of 1913, when Eliot's seminar had involved him in a continuing debate about evolution. This discussion in the 1913 seminar series ranged from the evolution of chemical properties to that of human ideas. The approach was a scientific one. Eliot chose material which relied heavily on anthropology, though the question had been raised as to whether Frazer, for example, could be called 'a scientific man'.[110] Eliot's paper was to ask the question, 'On what terms is a science of religion possible?' and to reach the conclusion that its author did not know.[111]

What its author stressed was that the phrase 'evolution of religion' was to be avoided. Eliot's paper contains a reminder to have, when criticizing as 'unscientific' the older definition of religion which said that it was striving towards a goal set by the historian, a 'note here against the "evolution" of religion'.[112] In the final version he criticizes the 'deplorable looseness' with which 'the word evolution is currently used', continuing,

The sorts of fact, as I understand it, which can properly be described in terms of evolution are those in which a continuous relation between organic tendency and environment can be expressed more or less quantitatively, according to a standard of value. We have the right to take human value as the standard for natural evolution, but what standard have we for religion or society?[113]

Eliot attacks the expression 'evolution of religion' by saying that natural evolution can be seen as a 'process from the point of view of *our* value, wh. is for our purposes conceived of as outside the process,' while on the other hand 'to some extent in a social progress, and to a very great extent in religious progress, the internal values are part of the external description'.[114] But he fails to explain what he means by an expression such as 'religious progress'. Does he simply mean change? Or does he mean an ascending development, that is, what we normally understand when we use the word evolution? Eliot himself does not confront this problem. But having attacked the phrase 'evolution of religion' at the start of his essay, we find

---

Gray, pp. 248–9 (*New Statesman* Webb review); I am grateful to Donald Gallup for alerting me to the *Saturday Westminster Gazette* review of Durkheim.

[110] See Grover Smith (1963), pp. 9, 67, and *passim*.
[111] Outline of IPR MS (King's College Library).   [112] IPR, p. 1 (verso).
[113] Ibid., p. 1.   [114] Ibid., p. 2.

him at its end troubled over the phrase 'evolution of mind'. Having
written this phrase, he has scored out the word 'evolution' and seems
to have substituted 'metamorphoses' before scoring that out too and
settling finally for the neutral term 'differences' which avoids any
problematical questions about advance or degeneration, so that the
chosen version of the sentence reads,

. . . Sociology and Comparative Religions, have a task as far as I know unique
among sciences: that of interpreting into one language an indefinite variety
of languages. History deals in one sense with a greater variety, in that it
deals with individuals—but these differences it is obliged to neglect: and
it is not concerned with differences of mind in so significant a sense as is
sociology.[115]

With regard to that first sentence Gray rightly says, 'here we have
the poet of *The Waste Land* speaking'.[116] But though materials
encountered in connection with this seminar paper would play their
part in the making of that poem, Eliot had not yet evolved the poetic
technique to handle them. Nor in this 1913 paper did he follow up
the suggestion that there was a common language into which the
different religions and their attitudes might be translated. Certainly,
though, the common language did not seem to be that of ordinary
speech, since to describe the actions of the savage in terms of modern
evolutionary doctrines, that is, to think of religion as 'a practical,
though imperfect or mistaken, adaptation to environment, more
or less consciously rational inventing of theories to account for
experience' was to mistakenly follow Tylor in a passage such as that
which Eliot quotes from *Primitive Culture*: 'The ancient savage
philosophers (sic) probably made their first step by the obvious
inference that every man has two things belonging to him, namely
a life and a phantom.'[117] The mistake here is to postulate 'in the
words of M. Lévy-Bruhl, the uniformity of mind'. Lévy-Bruhl in *Les
Fonctions mentales dans les sociétés inférieures* had shown the
pointlessness of modern thinkers' attempts to see the savage in their
own intellectual image, as, in that rather ridiculous phrase to which
Eliot has drawn attention with his '(sic)' in the above quotation from
Tylor, 'savage philosophers'. Yet while Eliot too objected to the
postulate 'uniformity of mind', he rejected what he saw as Lévy-
Bruhl's contention that the two ways of thinking, the prelogical and

[115] Ibid., p. 13.   [116] Gray, p. 133.
[117] IPR, p. 2A quoting Tylor, i. 428.

logical mentalities, were entirely different. Eliot wrote that Lévy-Bruhl 'appears to me to draw the distinction between primitive and civilized mental process altogether too clearly'.[118] This statement by Eliot is understandably evasive. He would repeat such an evasion in 1916, refusing to define just how much the savage and modern minds have in common.[119]

We should not take for granted the firmness of all the convictions which Eliot expresses in this paper, but it is valuable for its outline of problems that would continue to occupy his thought in the years that followed. That he himself could fall naturally into thinking in terms of a religious evolution is confirmed in a 1917 review of S. A. Cook's *The Study of Religions*. After complaining that Cook's work did not as the title led one to expect 'deal chiefly with primitive religion', he goes on to compliment Cook on having 'a great deal to say, and much that is extremely good, on the evolution of religion'. Significantly, Eliot's quotation reveals an interest in Cook's theory of 'survivals'. This anthropological term was used of strange features present within a culture which could not be explained in terms of any obvious current function. These were frequently explained in nineteenth- and early twentieth-century anthropology as 'survivals' of a previous culture which remained left over in the present day. Such an idea will be relevant for the examination of *The Waste Land*, but is also central to the quatrain poems where fragments of past texts and patterns of behaviour litter the present where they often seem uneasily out of place, though not eventually irrelevant. It is important that at the very time when Eliot was at work on these quatrain poems he noted Cook's theory that 'Survivals are not simply "left behind", they are subconsciously selected'.[120] We seem bound to carry on something of the Cyclops and of primitive rituals. The quatrain poems and those that follow, while not proposing 'uniformity of mind', appear to posit some link deeper than the safe world of the intellect. Writing of Stendhal and Flaubert as novelists who were 'men of far more than the common intensity of feeling, of passion', Eliot states that,

[118] IPR, pp. 3 and 5.
[119] Review of *Group Theories of Religion and the Religion of the Individual*, *International Journal of Ethics*, Oct. 1916, p. 116.
[120] Review of Stanley A. Cook, *The Study of Religions*, *Monist*, July 1917, p. 480.

The surface of existence coagulates into lumps which look like important simple feelings, which are identified by names as feelings, which the patient analyst disintegrates into more complex and trifling, but ultimately, if he goes far enough, into various canalizations of something again simple, terrible and unknown.[121]

The ghost of Bergson may haunt those words, but they were also the words of the man who had written a short while before that 'The artist, I believe, is more *primitive*, as well as more civilized, than his contemporaries, his experience is deeper than civilization, and he only uses the phenomena of civilization in expressing it.'[122] Literature could be both a humiliating and a bloodthirsty business. 'Stendhal's scenes, some of them, and some of his phrases, read like cutting one's own throat.'[123] The artist (and the accurate reader) then is in an ideal position to appreciate the savage, as well as being distanced from him by being on the far side of contemporary civilization. He is at once closer to and further from the savage, another reason for Eliot's linking of 'the poet and the anthropologist', both of whom are in the business of comprehending 'the stratifications of history that cover savagery'.[124]

Such an artist is a kind of Bolo: the sophisticated man-about-town, yet also one who has resort to the primitive world of Bolovians and the potency of a Big Black Queen. He is the Baudelairean dandy, the Mr Eliot of the pin-stripe suit, but the ritual world of the savage was also available to him, as Baudelaire and the Symbolists had indicated. Such a view was not Eliot's alone. Pound in 1914 was writing of artists as 'the heirs of the witch-doctor and the voodoo', but however much Eliot was fascinated by the frenzy of the primitive, he also approached it with some detached scientific interest.[125]

In the 1913 paper we can see Eliot's eye lighting on various specific rites that would be of importance to him. Foremost among these is the 'death and revival of vegetation' which Frazer associates with numerous sexual rituals.[126] These vegetation ceremonies, passed down through generations, became linked to Eliot's examination of

---

[121] 'Beyle and Balzac', *Athenaeum*, 30 May 1919, p. 393.
[122] 'Tarr', *Egoist*, Sept. 1918, p. 106.
[123] 'Beyle and Balzac' (C80), p. 393.
[124] 'War-paint and Feathers' (C94), p. 1036.
[125] Pound, 'The New Sculpture', *Egoist*, 16 Feb. 1914, p. 68.
[126] J. G. Frazer, *The Golden Bough, Third Edition, Part III, The Dying God* (London: Macmillan, 1911), p. 263.

tradition. The primitive connection between sexuality, plant-life, and religion joined his interest in primitive rites to his interest in the religio-sexual frenzy of the mystics. If modern discoveries and theories in eugenics and philosophy connected heredity in plants to questions of better human breeding, this afforded a striking ironic link with primitive man, a link which Eliot would come to exploit with peculiar poetic richness. He would return to Frazer's work at several critical moments in his writing. In 1913 Eliot both praises and attacks Frazer.

No one has done more to make manifest the similarities and identities underlying the customs of races very remote in every way from each other. Just as through the historical mode of research, certain fixed relations can be found which are not relative to the observer; but the nature of the entities between which the relations obtain is never completely known. I have not the smallest competence to criticize Dr. Frazer's erudition, and his ability to manipulate this erudition I can only admire. But I cannot subscribe for instance to the *interpretation* with which he ends his volume on the Dying God. He is accounting for the magical rites of spring festivals: — [127]

Eliot then gives the reference 'P. 266' which refers to Frazer's expla-nation of 'The Magic Spring' in terms of 'the quickening and fertilising influence which the spirit of vegetation is believed to exert upon the life of man as well as of plants'. Frazer went on to talk of spring, describing how a modern man could imagine himself as a savage misguided as to 'the true causes of things' and so thinking that 'by masquerading in leaves and flowers he helped the bare earth to clothe herself with verdure'.[128] Eliot objected to this since it assumed, as Tylor had assumed, a 'savage philosopher'. He was more attracted to the theory of a 'prelogical mentality' put forward by Lévy-Bruhl who saw the savage as involved in a mystical 'law of participation' connecting him with his environment, in a manner alien to the framework of modern logical thought. The savage therefore was not simply 'wrong' as Frazer would have it. Even then, though, Eliot argues, an act of falsifying interpretation would be needed to relate the modern to the primitive mentality.[129] The essay reaches a philosophical impasse. Yet, for Eliot's poetry it signalled the way ahead. The conclusion of *The Dying God* to which Eliot referred discussed Mannhardt's analysis of European vegetation rites which

. . . fallen from their high estate, no longer regarded as solemn rites on the punctual performance of which the welfare and even the life of the

[127] IPR, p. 11.   [128] Frazer, pp. 266–7.   [129] IPR, pp. 11–13.

community depend . . . sink gradually to the level of simple pageants, mummeries, and pastimes, till in the final stage of degeneration they are wholly abandoned by older people, and, from having once been the most serious occupation of the sage, become at last the idle sport of children.[130]

Such ideas reinforced not only contemporary speculations about the origins of drama in religion; they also tied in with Eliot's view of a modern world as run-down and 'refined beyond the point of civilization'. Frazer's concluding discussion of Australian ceremonies paved the way for Eliot's much more detailed reading of Durkheim on the same.

In Durkheim's *Elementary Forms of the Religious Life* it was the factual, anthropological element which most caught Eliot's interest, though he thought that 'the whole book is intensely interesting'.[131] This work by Durkheim relies very heavily on the anthropological fieldwork of Spencer and Gillen and that of Howitt. These anthropologists had worked on the Australian aborigines who had fascinated the anthropological writers whom Eliot had read before. Like Lang and Frazer, Tylor had placed aborigines 'among the lowest living men'. Jane Harrison had referred to their *intichiuma* ceremonies with regard to 'the renewed fertility of the earth'. She saw the ceremony as a *methexis* (participation) which preceded the development of *mimesis*.[133] Lévy-Bruhl too had commented on this rite's utilitarian and mystic significance.[134] But it was Durkheim who, again treating the aborigines as possessors of 'the most primitive and simple religion which is actually known', provided most detail about and gave the greatest emphasis to these fertility rites, particularly the *intichiuma* ceremonies of the Witchetty Grub clan of the Arunta tribe, performed to secure the fertility of animals and plants.[135] These ceremonies were followed by the consumption of the totem animal, a ceremony for which Durkheim uses the word 'communion' and which he relates to Robertson Smith's work on Hebrew sacrificial ceremonies and

[130] Frazer, p. 269.
[131] Review of Durkheim, *Elementary Forms*, *Monist*, Jan. 1918, p. 158.
[132] Tylor, i. p. 242.   [133] Harrison, *Themis*, pp. xiv and 125–6.
[134] Lucien Lévy-Bruhl, *How Natives Think* (authorized trans. of *Les Fonctions mentales dans les sociétés inférieures*, trans. by Lilian A. Clare, London: Allen & Unwin, 1926), pp. 91 and 250.
[135] Emile Durkheim, *The Elementary Forms of the Religious Life*, trans. by Joseph Ward Swain (London: Allen & Unwin, 1915; seventh impression, 1971), p. 1.

the way in which 'sacrificial banquets have the object of making the worshipper and his god communicate in the same flesh'.[136] Durkheim explains that, 'It has been shown how a whole series of preliminary operations, lustrations, unctions, prayers, etc., transform the animal to be immolated into a sacred thing, whose sacredness is subsequently transferred to the worshipper who eats it.' Such animals contain a mystic substance which forms part of the soul of the member of the totem; the growth of the animal or plant and of the man are thus inextricably related, so that there are both rites of communion and related 'rites destined to assure the fecundity of the animal or vegetable species which serves the clan as totem'. The extremely primitive nature of these ceremonies is stressed. 'These rites are certainly among the most primitive that have ever been observed. No determined mythical personality appears in them; there is no question of gods or spirits that are properly so called; it is only vaguely anonymous and impersonal forces which they put into action.'[137] Re-reading Durkheim's *Elementary Forms* Eliot recalled his earlier anthropological reading, stating that Durkheim's book succeeded most in 'the purely anthropological aspect', and arguing that its views had to be considered with and compared to those of previous interpreters like Tylor, Max Müller, Andrew Lang, Jevons, Robertson Smith, and Mannhardt.[138] But Eliot no longer agonizes over what constitutes a 'fact'. Rather he continues to wish to avoid what he had called in his seminar paper draft 'examples of overinterpretation'.[139] Eliot's view was that Durkheim was most persuasive when he was most factual and avoided metaphysical speculation or when he concentrated on arguing against his predecessors. 'M. Durkheim's theory [of Australian totemism] is the best because it is the nearest to being no theory at all. And when he comes to state it in positive terms, he finds almost as much difficulty as his predecessors in avoiding intellectualization.'[140] This avoidance of 'intellectualization' is seen as a goal. While Eliot finds Durkheim's theory 'a contribution', he worries about whether it is 'capable of articulate expression'. Such concerns had long been part of Eliot's own poetry—

It is impossible to say just what I mean!

—but by the time of the Durkheim review such an outcry has acquired a particularly religious significance:

---

[136] Durkheim (1971), p. 337. [137] Ibid., pp. 337 and 341. [138] As n. 131 above.
[139] Outline of IPR, p. 2 (King's College).   [140] As n. 131 above.

Signs are taken for wonders. 'We would see a sign!'
The word within a word, unable to speak a word,
Swaddled with darkness.

That word 'swaddled' with its suggestion of Christ's 'swaddling
clothes' (Luke 2: 12) places us firmly within the area of fertility and
birth. In these lines from 'Gerontion' we witness the birth of a god
in the manner suggested by Harrison. The god is born from the choric
cry. Gerontion longs for emotions. The relationship between Eliot's
anthropological reading and his poetry of this period will be explored
further in the following chapter. For the moment we may note that
his own personal circumstances sometimes find their way into Eliot's
remarks on the 'savage'. In 1916, after giving up an exhausting
teaching job, before becoming a bank clerk, and while splitting his
time between reviews, lecturing, looking for a job, and writing
strangely intense 'impersonal' poems, Eliot's interest turned to the
way in which 'the savage lives in two worlds, the one commonplace,
practical, a world of drudgery, the other sacred, intense, a world
into which he escapes at regular intervals, a world in which he is
released from the fetters of individuality'. He had been reading
about Durkheim's observations and conclusions arising from his
consideration of the facts of primitive religion. Eliot reveals his
awareness that these were based largely upon the researches of
Spencer and Gillen, and Howitt, into Aboriginal society, its totemic
beliefs, and its periodic festivals. Eliot saw Durkheim as the important
leader of a movement that had grown in the last ten years and which
was reinterpreting more than just primitive religion. He draws
attention to the way in which Parisian writers such as Lévy-Bruhl,
Hubert, Mauss, Hamelin, and others, as well as English scholars
including Jane Harrison, Cornford, and A. B. Cook 'bear witness
to the fertility of Professor Durkheim's ideas. His present volume,
well translated, though with less literary finish than the original,
ought to be read not only by specialists, but by everyone who is
interested in the history and in the future of religion.'[141] That
submerged pun 'the fertility of Professor Durkheim's ideas' shows
Eliot's creative irony at work. The sophisticated thought of the
anthropologists is seen as itself a sort of fertility ceremony, the
sophisticated paralleling the savage. Certainly Eliot the London

[141] 'Durkheim', *Saturday Westminster Gazette*, 19 Aug. 1916, p. 14.

banker saw the savage as important to modern man and particularly to the artist. Anthropology, imaginatively used, could further the new art. As he wrote in 1919, aiming, no doubt, to disconcert over-comfortable readers of the *Athenaeum*, it is not enough for modern man to have some knowledge of Freud and Fabre, it is also essential to know a bit about

the medicine man and his works. Not necessary, perhaps not even desirable, to know all the theories about him, to peruse all the works of Miss Harrison, Cooke, Rendel Harris, Lévy-Bruhl or Durkheim. But one ought, surely, to have read at least one book such as those of Spencer and Gillen on the Australians or Codrington on the Melanesians. And as it is certain that some study of primitive man furthers our understanding of civilized man, so it is certain that primitive art and poetry help our understanding of civilized art and poetry. Primitive art and poetry can even, through the studies and experiments of the artist or poet, revivify the contemporary activities . . . More intelligibly put, it is that the poet should know everything that has been accomplished in poetry (accomplished, not merely produced) since its beginnings—in order to know what he is doing himself. He should be aware of all the metamorphoses of poetry that illustrate the stratifications of history that cover savagery. For the artist is, in an impersonal sense, the most conscious of men; he is therefore the most and the least civilized and civilizable; he is the most competent to understand both civilized and primitive.

Consequently, he is the most ready and the most able of men to learn from the savage . . .

Eliot links the artist with the savage; but here rejects the naïve idea that the savage has gifts of mystic perception or artistic sensibility not possessed by the artist.[142] Eliot would waver on this point. In 1916, though he would criticize Lévy-Bruhl for exaggerating the difference between the mind of the savage and the mind of a civilized man, he would none the less be attracted to the idea that the mystical mentality, albeit at a low level, played a much greater part in a savage's daily life than in that of a civilized man.[143] What is clear is that, though his ideas of the precise nature of the relationship fluctuated, Eliot continued to relate primitive to civilized man in the period leading up to 1920. In that year he would ironically relate the evolution of anthropological and related theories to the evolution of the fertility ceremonies with which such scholarship dealt.

---

[142] 'War-paint and Feathers' (C94), p. 1036.   [143] As n. 119 above.

Few books are more fascinating than those of Miss Harrison, Mr. Cornford, or Mr. Cooke, when they burrow in the origins of Greek myths and rites; M. Durkheim, with his social consciousness, and M. Lévy-Bruhl, with his Bororo Indians who convince themselves that they are parroquets, are delightful writers. A number of sciences have sprung up in an almost tropical exuberance which undoubtedly excites our admiration, and the garden, not unnaturally, has come to resemble a jungle. Such men as Tylor, and Robertson Smith, and Wilhelm Wundt, who early fertilized the soil, would hardly recognize the resulting vegetation; and indeed poor Wundt's *Völkerpsychologie* was a musty relic before it was translated.[144]

Eliot's concern with anthropology at this time is like his concern with religion. He both finds it important and mocks it. Reviewing Wundt's 'musty relic' in 1917, he had complained in terms quite contrary to those of his 1913 paper, 'We find nothing of the influence of the sexual instinct, for instance, upon religion and myth. Mysticism is not even included in the index. The treatment of primitive art quite neglects its aesthetic value . . . why, among much matter about religious cults, is there so little about religious feelings? It is with the external features of development that Wundt is concerned.'[145] Later, in a reworking of this review, Eliot makes it quite clear that he feels that the sexual aspect of religion stressed by anthropologists such as Durkheim is important not only in the context of the primitive. 'And of the role which the sexual instinct plays in the religion and mythology of primitive peoples (indeed all religion) Wundt has almost nothing to say.'[146] Eliot's eagerness to employ anthropological findings to religion in general was not his only concern with anthropology. He also found its imagery relevant in writing of his personal life. Plagued by his own sexual anxiety, a frequent theme of the poetry, he wrote to Aiken in 1914 of how he was writhing in impotence in Oxford.[147] His attitude towards this town resembled his attitude to Boston, and he longed for a Gauguin-like existence, more primitive, Bolo-esque life: 'Come, let us desert our wives, and fly to a land where there are no Medici prints, nothing but concubinage and conversation . . . Oxford is very

[144] *SE*, pp. 62–3.
[145] Review of Wilhelm Wundt, *Elements of Folk Psychology, International Journal of Ethics*, Jan. 1917, pp. 253–4.
[146] Review of Wilhelm Wundt, *Elements of Folk Psychology, Monist*, Jan. 1918, p. 160.
[147] Letter to Aiken, 30 Sept. [1914] (Huntington).

pretty, but I don't like to be dead.'[148] When he wrote this to Aiken,
Eliot was unmarried. Soon marriage and its effects would agonize
him. He feared in 1915 that his marriage might alienate him from his
family, that the values which seemed to him increasingly important
(values of religious, literary, sexual, and geographical preference)
might be judged harshly by history, and that his life might be
seen as a failure. In February 1915 he examined *The Origin and
Development of the Moral Ideas* by Edward Westermarck, an
anthropologist hailed as 'the Darwin of Moral Science'.[149] One of
Eliot's 1915 papers links the Polynesians and Anglo-Saxons and
worries that future moralists may deny the 'rightness' of present
actions. One can see under his consideration of Westermarck's
book Eliot's worry that the motives for his own action may be
incomprehensible.

We can trace the origin and development of the moral ideas. It is the business
of descriptive ethics to follow the slow expansion of primitive desires into
the various systems of values, indicating the bypaths of prejudice and the
blind alleys of taboo and superstition, noting the categories or the general
form into which value articulates itself; the various genres of art, the various
satisfactions of religion, the subordinate moral values, as benevolence,
sympathy, or self-sacrifice. It can show too the causes for the assertion of
particular virtues and vices at particular times, the adaptation to particular
needs, or the submission to particular compulsions. It can show the gradual
approximation to an identical morality for all men. But it cannot explain
the meaning of the sort of thrill which I have at the sight of a new and
satisfying work of art, or a perfect response to a new moral situation; and
this thrill is the intrinsic value. The next moment may show that I was wrong;
time which, as the Hindoo Sage observes, is hard to beat, may reconstruct
every value; but at the moment and for that consciousness, the value was
there, and remains inexplicable.[150]

Appropriately in terms of his attraction to Buddhism, and in terms
of *The Cocktail Party*, Eliot's essay ends with a consideration of the

---

[148] Letter to Aiken, 31 Dec. 1914 (Huntington), quoted in slightly abbreviated
form in Aiken, 'King Bolo and Others', pp. 22–3.

[149] See Robert Crawford, 'T. S. Eliot, Lawrence of Arabia, and Oxford Anthropology, 1914–15', *Journal of the Anthroplogical Society of Oxford*, Hilary 1984,
pp. 45–54.

[150] 'Ethics', undated Oxford TS in Houghton Library, pp. 11–12. I date this
21-page essay to *c*. March 1915 because of the reference to Westermarck's book which
Eliot borrowed at the end of February.

injunction to 'work out your salvation with diligence'.[151] But the writer is unsure about what to make of such an instruction. What Eliot was sure of was that he was unsettled and passing through a period of personal crisis. Writing to Aiken in 1914 he cloaks his personal position in Frazerian language, stating that it is worth while from time to time tearing oneself to pieces and waiting to see if the bits will sprout.[152] Later, the image would reappear, made more impersonal, in *The Waste Land*. There is much covert autobiography in the early Eliot. In a 1917 short story his Eeldrop and Appleplex seek out a dingy neighbourhood in 'a disreputable part of town', and observe the crowds in the street questioning the onlookers, while recording their manners and turns of phrase.[153] They behave like anthropologists. It is perhaps more significant, though, that in 1916, the Eliot who had given up America and philosophy for England and a struggle to make his name in literary journalism should see himself pulled to pieces by various identities,

> En Amérique, professeur;
> En Angleterre, journaliste; . . .

Some of these are the urbane, urban poses of London where the speaker is 'un peu banquier', or Paris, where he is sophisticatedly unconcerned. Others, though, look with irony to a very different world, but one which would be crucial for his poetry.

> Je célébrai mon jour de fête
> Dans une oasis d'Afrique
> Vêtu d'une peau de girafe.
>
> On montrera mon cénotaphe
> Aux côtes brûlantes de Mozambique.

Attending a 1916 meeting of the Aristotelian Society along with Ezra Pound, Eliot explained that his friend was present only as an anthropologist.[154] But Eliot too was observing with an anthropological eye both the manners of the society around him and his own actions. As a poet his observations were at once more concrete and more imaginatively disturbing than those of any anthropologist

---

[151] *CPP*, p. 411.   [152] As n. 147 above.
[153] 'Eeldrop and Appleplex', *Little Review*, May 1917, p. 7.
[154] See Pound (1950), p. 331 (letter to Santayana 8 Dec. 1939).

or dealer in philosophical abstractions. He found the city clerk and the primitive man in the one body and psyche. From the 'Gold Coast' of Harvard, Eliot had brought the savage to the City of London.[155]

---

[155] The 'Gold Coast' was the name of the expensive area in Cambridge where Eliot lodged when a Harvard undergraduate; see Gordon, p. 20.

IV

WAITING FOR RAIN

ELIOT was always a careful listener. Some phrases lay in his mind for years. Others were more quickly put to use. On 29 October 1914, he heard Collingwood, lecturing on Aristotle's *de Anima*, talk of how the soul might be supposed to leave the body and return to it later. Eliot scribbled a reference to Frazer's treatment of such experiences in *The Golden Bough*.[1] By February 1915, he was sending Aiken 'Suppressed Complex', a poem about a soul-like shadow dancing in the firelight of a woman's room, before passing out of the window.[2] In 1916 Eliot recalled 'Tylor's dreaming aborigine who finds that his soul in sleep can part company with his body and roam the forests . . . '.[3] In 'The Hollow Men', the concept of wandering souls would link anthropology and the romantic world of Dowson with his

> hollow lands
> Where the poor, dead people stray,
> Ghostly, pitiful and gray.[4]

The transfer of souls in reincarnation would be important for *The Waste Land*, but in 1915 the 'revitalizing of the classics', as Eliot later described one of the effects of anthropology, seems most apparent in 'Mr. Apollinax' where we are made very aware of the primitive, and particularly sexual aspects of classical mythology.[5] In the few years that followed, Eliot adopted various strategies to keep his poetry flowing. In each the primitive, sometimes bestial is joined obdurately to the modern and sophisticated. City men walk by the Thames like wolves in 'Le Directeur'; 'Mélange Adultère de Tout' links London banker and giraffe-headed African. Part of the reason for this is no doubt Eliot's hostility to ideas about 'social progress' one aspect of which was the eugenics movement. Reviewing

---

[1] '[Notes on Aristotle]' (Houghton).
[2] Letter to Aiken, 7 Feb. 1915 (Huntington).
[3] 'Durkheim.'   [4] Dowson, 'A Requiem', *Poetical Works*, p. 64.
[5] 'A Prediction in Regard to Three English Authors', *Vanity Fair*, Feb. 1924, p. 29.

McBride, Eliot quoted from that writer's conclusion that, given proper education of mind and body, 'the next generation may start at a very slightly higher level of capacity than their fathers'.[6] The quatrain poems give the lie to that. Also important is his sympathy for 'the classicist point of view' which, he stated in 1916 'has been defined as essentially a belief in Original Sin'.[7] The quatrain poems bind up such sympathies with a way of thinking which owed much to anthropology and Eliot's growing wish to include in his work the worlds of both the savage and the city.

In 1918 Eliot wrote of the need for artists to be at once very primitive and very sophisticated; he praised Lewis's *Tarr* as being 'like our civilization criticized, our acrobatics animadverted upon adversely, by an orang-outang of genius, Tarzan of the Apes'.[8] Eliot's Sweeney fulfilled this function. A caricature of Dr Sweany's 'manly' man, he is a modern troglodyte, but one with a rich ancestry. Book IX was by far the most thumbed section of Eliot's *Odyssey*, and it was the story of the Cyclops that 'savage man' which he read most attentively.[9] Though he found the passionate irresponsibility of the Homeric world 'shocking', preferring its Virgilian successor 'because it was a more civilized world of dignity, reason and order',[10] Eliot also realized that the all-important 'mind of Europe' was one 'which changes, and that this change is a development which abandons nothing *en route*, which does not superannuate either Shakespeare, or Homer, or the rock drawing of the Magdalenian draughtsmen'.[11] It contained not only the Homeric, Cyclops and all, but also the troglodytic itself with its own representations, the dawn of the mind. The Cyclops enters Eliot's poetry after a line which seems to mock the rising of the formulaic Homeric 'rosy-fingered dawn' in the poem 'Sweeney Erect' when:

> Morning stirs the feet and hands
> (Nausicaa and Polypheme).

It dawns on us that we are back at temporal beginnings and slipping down the evolutionary tree. Refusing to let even this distorted version

[6] 'Recent British Periodical Literature in Ethics', *International Journal of Ethics*, Jan. 1918, p. 274.

[7] 'Syllabus of a Course of Six Lectures on Modern French Literature' (1916), rpt. Moody, p. 44.

[8] 'Contemporanea' (C64), p. 84.

[9] Eliot's copy of the *Odyssey* is in King's College Library.

[10] *OPP*, p. 124.   [11] *SE*, p. 16.

of myth lie still, the next lines confirm the evolutionary idea, lowering its level while at the same time raising the domestic temperature to the tropical:

> Gesture of orang-outang
> Rises from the sheets in steam.

Seedy modern guest house or brothel becomes primitive jungle. Somewhere in between is a meeting on a Greek island. Sweeney's encounter with 'the epileptic on the bed' parallels that of Nausicaa not with Odysseus but, more shockingly, with 'Polypheme'. Sweeney's name pronounces him of the lineage of a man who sold human flesh for butchermeat, a 'demon barber'. Eliot traces the line further back into primitive prehistory by an explicatory aligning of his protagonist and a putative ancestor, the Cyclops, who devoured Odysseus's comrades. When Nausicaa meets Polypheme beauty meets the beast. It is also the meeting of the cultured city dweller, who led Odysseus from undergrowth to city palace, with the savage Cyclops who wished to eat the hero in a cave. The absurd encounter represented in Eliot's parenthesis is erotic, but also cannibal, each aspect reinforcing the outrageous horror-comedy of the other. The situation and its elements foreshadow the later confrontation of Sweeney and Doris on their 'cannibal isle'. The paralleling of Sweeney and the Cyclops here does on a small scale the work of the anthropologists' comparative method, 'to make manifest the similarities and identities underlying the customs of races very remote in every way from each other', though both are part of the 'mind of Europe'.[12] The artifice of the opening stanzas with their

> Paint me . . .
> Paint me . . .
> Display me . . .

sets us at a painterly remove from decorously observed suffering and does little to prepare us for the replacement of the distant tangled hair of Ariadne by immediate presence of the directly primitive

> This withered root of knots of hair
> Slitted below and gashed with eyes . . .

[12] IPR, p. 11.

The Classical gives way to the savage and language grows hyperactive. As the poem progresses it is beast, not beauty, which meets beast. 'Root' is normally associated with a single hair, not 'knots' of it. 'Root' here links the figure in the sheets to a tree from which life has drained, yet soon it will come alive with galvanic motion. The resurrection of dead life through the repetition of supposed past actions will be a theme of the poem, and the poem will also be about suggesting a 'root' for primitive (which here includes ape-like primitive) behaviour through mythological explanation which involves the merciless demystification of myth. Eliot has learned from his anthropology.

Stanza three's shockingly different mythology prepares us for 'This withered root'. When Nausicaa meets Polypheme mythology is cut up and reassembled to offer a shocking explanation of what underlies it, a parenthetical explanation which throws a provoking light on the relations between men and women, between civilization and savagery, between present and past. The second half of the verse, 'Gesture of orang-outang / Rises from the sheets in steam' simultaneously forces us back into human prehistory, before even Polyphemus, and forward into the present of Doris and Mrs Turner, since monkey evolves into human where 'orang-outang' becomes *homo erectus*; yet 'knots of hair' makes the modern return to the ape, and shaving Sweeney, for all his performing the action of his namesake, the demon barber, seems as crude and brutal as the Cyclops—such is the grotesquely comic evolutionary irony of the loss of hair. The poem lets us posit an evolutionary series,

Orang-outang
↓
Cyclops
↓
Sweeney

but the way in which these three are introduced and relate to one another upsets any simple progressive development by raising questions about what evolutionists and anthropologists discussed as 'reversion' and 'degeneration', countering all optimistic ideas about the growth of humanity. Sweeney has no sooner stood up and shaved than he appears in a vision of reversion taking the form of the second parenthesis, the importance of whose material strains against the parenthetical format, guiding our reading of the poem:

(The lengthened shadow of a man
   Is history, said Emerson
Who had not seen the silhouette
   Of Sweeney straddled in the sun.)

Sweeney is a weapon against a too hopeful view of history. Emerson's essay on 'history' begins, 'There is one mind common to all individual men', whereas Eliot's anthropological reading had taught him to be wary of the nineteenth-century assumption of what was 'in the words of M. Lévy-Bruhl, the uniformity of mind'.[13] As poet, Eliot makes devastating use of such an assumed uniformity. Inverting the 'savage philosopher' — a savage who attempts to reason like a modern — Eliot produces a modern who behaves like a savage, shockingly contradicting the idea of progress and modern 'self-reliance'. Emerson wrote that he could find 'the primeval world . . . in myself', but his thinking self is seen as confident master of that world.[14] As a commentary on such sentiments Eliot's poem reads with a vengeance. His criticism of Lévy-Bruhl for overstressing the distance between the savage and the modern mind shows that for Eliot the two were linked, but Lévy-Bruhl's stress on the different, apparently unreasonable nature of the savage leads to the modern Sweeney whose world is far removed from the sweet reason of Emerson. Sweeney shocks because he shows so clearly the effect of lack of *themis*. Harrison had written of such a lack among the Cyclopes. In *Themis* she stressed the need for group rather than individual values in conducting life and saw them as the foundation of religion. Sweeney cannot be outgrown, but must be controlled. Eliot, however, is content to show this here without curbing Sweeney's activities; they are too useful as violently funny weapons against the high priests of the progressive individualism of Eliot's youth, whom he saw as false prophets of a rigid pseudo-themis, 'Matthew and Waldo, guardians of the faith, / The army of unalterable law.' Sweeney stands against those whose personal taste is too closely linked to morality in a falsely genteel way. Himself unspeaking, armed with his razor he reduces life to the level of a primitive 'shriek' with which 'the ladies of the corridor', despite the way in which they 'Call witness to their principles / And deprecate the lack of taste', find themselves inescapably 'involved'. Just as there is a tension between the

---

[13] Emerson, *Works* (London: Routledge, 1895), p. 1; IPR, p. 3.
[14] Emerson, p. 3.

evolutionary ups and downs of the hairy epileptic on the bed who
'Jacknifes upward', and the shaving 'Sweeney Erect', so there is set
against the prostrate unreason indicated by 'The epileptic on the bed
[who] / Curves backward, clutching at her sides' another woman
who later walks upright:

> But Doris, towelled from the bath,
>   Enters padding on broad feet,
> Bringing sal volatile
> And a glass of brandy neat.

Arriving like a final *dea ex machina*, Doris's condition makes her
like an ironic version of that goddess in Tennyson's 'Oenone', 'Idalian
Aphrodite beautiful, / Fresh as the foam, new-bathed in Paphian
wells.' Certainly there is more mythology behind the poem.

Moody thinks that 'the final stanza may allude to a variant of
Ariadne's tale, which has it that she did not die of a broken heart
but was loved by Bacchus'.[15] The relevance of this variant comes
earlier in the poem whose opening stanzas call for a painting of
Ariadne deserted on Naxos. What follows connects Sweeney both
with his mate, an 'orang-outang', and with a god, 'Polypheme' being
Poseidon's son. Moreover, Sweeney is an ironic Bacchus to his own
Ariadne. Eliot had written an earlier 'Debate between Body and Soul',
entitled 'Bacchus and Ariadne', and probably inspired by Titian's
painting in the National Gallery.[16] He ticked this picture in his
London *Baedeker* whose commentary singles it out for its great
'exuberance'.[17] The 'exuberance' of Sweeney forms a disturbing
parallel to that of the god, since it so obviously belongs to what
Aspatia in *The Maid's Tragedy* (from which Eliot's epigraph comes)
calls 'that beast man'.[18]

Anthropology for Eliot did not remake the myths, but showed how
they had, while becoming the possessions of high culture, transmitted
and not entirely transmuted primitive origins. By purporting to de-
interpret them by removing the excrescences of later interpretations
it made possible a reinterpretation which allowed mythology to be
seen again as something that while still existing on the level of the

---

[15] Moody, p. 61.   [16] Miscellaneous Material MS (Berg).
[17] Karl Baedeker, *London and its Environs* (Leipzig: Dulau & Co., 1908),
p. 175; references are to Eliot's copy (dated 1910) in King's College Library.
[18] Beaumont and Fletcher, *The Maid's Tragedy*, ed. Andrew Gurr (1619; rpt.
Edinburgh: Oliver & Boyd, 1969), II. ii. 30.

most civilized and polished communication kept speaking of what it had sprung from—men's basic needs and desires. Yet the very study of mythology from a scientific attitude distanced modern man at the same time as it seemed to bring him closer. The size of the gap between students and studied fluctuated. In 'Sweeney Erect' the modern seems closer to the savage Polyphemus than to Emerson, and the delight in attacking the nineteenth century is obvious; but so is the tension between modern and ancient. It is not just that high myth looks down towards low present reality. If Sweeney is a Bacchus, the god himself is nowhere present in the poem. Theseus's ship, Aeolus, and the abandoned Ariadne—these too are absent, or at least they are present only in some later wish for an imaginative reconstruction,

> Paint me . . .
>
> Paint me . . .
>
> Display me . . .

'Sweeney Erect' is a poem about sexuality; that sort of erection clearly links at the most basic level the different planes of history in the poem. But it is also about the unstable evolution both of *homo erectus* and his culture, all of which may be a vast illusion if what man does is simply continue his savagery while trying to repeat faint echoes of some suspect original grandeur. The movement from the initial grandly imperative wish for a creative act, reviving an older myth, to the final mundane narrative of the beginning of another, much less magnificent revival of potential creativity promised by 'sal volatile / And a glass of brandy neat' is a movement away from a first situation (that of Ariadne on Naxos) which we never see in itself; the painting conjured up and the other parallels to this first situation are interpretations not just of each other, but also of that first situation which, because a 'myth' and so subject to constant reinterpretation, may never have happened in any of the ways presented, if indeed it ever took place at all. Eliot had hinted in his primitive ritual paper that grand structures of belief might be founded on something false or grotesquely crude or even meaningless. This idea grows in the poems leading up to 'The Hollow Men', finding full expression in 1923 when he looks back to his seminar paper, speculating that 'primitive man' may have

acted in a certain way and then found a reason for it. An unoccupied person, finding a drum, may be seized with a desire to beat it; but unless he is an

imbecile he will be unable to continue beating it, and thereby satisfying a
need (rather than a 'desire'), without finding a reason for so doing. The reason
may be the long continued drought. The next generation or the next
civilization will find a more plausible reason for beating a drum. Shakespeare
and Racine—or rather the developments which led up to them—each found
his own reason. The reasons may be divided into tragedy and comedy. We
still have similar reasons, but we have lost the drum.[19]

Here Shakespearian splendour, and indeed all art, is seen as springing
from something beyond reason, and possibly senseless. The Eliot
who underlined Plotinus's φεύγωμεν δὴ φίλην ἐς πατρίδα (finding the
same exhortation to 'return to the beloved fatherland' in an essay
which he translated in 1927) and who wrote *Wanna Go Home,
Baby?* was always concerned with returning to sources.[20] But in this
return he sometimes found something as horribly inane as the tedium
of the city clerk's world. The quatrain poems identify in man lowest
common denominators operating below the level of any 'savage
philosopher's' thought. Their characters, like those of Jonson's
comedies, are grotesquely stripped down and caricatured. 'This
stripping', Eliot wrote, 'is essential to the art.'[21] We recognize in the
quatrain poems what Eliot occasionally found in Davidson: that
though his 'dwelling-place' alters, the human 'tenant . . . is the same
through all ages', but Eliot's linking factors, though they may be
comic and self-mocking, are scarcely reassuring.[22] The various
sexual liaisons which parallel one another in 'Sweeney Erect' are like
Jane Harrison's ritual *dromena*: things re-done and pre-done.
    In 'Burbank with a Baedeker: Bleistein with a Cigar' we have
another central situation whose repetition makes it analogous to one
of Harrison's rituals. The situation is that of the outsider meeting
the pleasures of a different, reputedly splendid civilization. In the
central *dromenon* these pleasures take an explicitly sexual form,
represented by Burbank's encounter with Princess Volupine. But such
a situation is also presented as a thing pre-done, interpreted by being
seen in terms of Antony's meeting with Cleopatra as portrayed by

---

[19] 'The Beating of a Drum' (C146), p. 12.
[20] Eliot's copy of *Plotini Enneades* (King's College Library); the Greek phrase is
part of the epigraph to Jacques Maritain 'Poetry and Religion [I]', which Eliot
translated for *Criterion*, Jan. 1927, p. 7; 'Fragment of an Agon. From *Wanna Go
Home, Baby?*', *Criterion*, Jan. 1927, p. 74.
[21] 'The Comedy of Humours', *Athenaeum*, 14 Nov. 1919, p. 1181.
[22] 'Reflections on Contemporary Poetry', *Egoist*, Oct. 1917, p. 134.

Shakespeare, and as a thing re-done, since the third meeting, that of Princess Volupine with Sir Ferdinand Klein presents us again with a similar situation. So we have the basic pattern of mutual interpretation, each performance of the ritual interpreting the others, which Eliot had examined in 1913. It is convenient to relate the three meetings in terms of a diagram similar to that used by Eliot in his paper.[23] (See Figure 3.) We are not told that these are part of a continuing ritual, but we are encouraged to treat them as such by the similarity present in the three meetings which are connected with

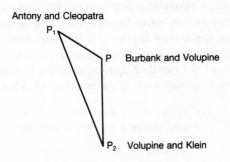

Fig. 3. *Diagram in Eliot's Paper*

each other further by the Shakespearian language applied to each. It is a poem about cultural change. By using such phrases as

> the God Hercules
> Had left him, that had loved him well.

and

> Her shuttered barge
> Burned on the water all the day.

and

> On the Rialto once.

and

> Lights, lights,

[23] IPR, p. 10.

Eliot is able to recall Shakespeare closely enough to suggest at first
sight that an ironic contrast is aimed at, but also to present the
possibility that the allusion functions as the poetic equivalent of a
legal fiction. The words stay the same, but we are forced to ask
awkward questions about how much their meaning has altered. The
simple ironic reading is based on the assumption that the high
Shakespearian allusions are really inapposite. The other, more
disturbing reading is that the Shakespearian phrases are made
appropriate because the present situation reinterprets them in such
a way that we are forced to wonder if a Burbank–Volupine situation
was not just the sort of thing they were talking about all along. This
use of Shakespeare, the central figure of Anglo-Saxon culture, stresses
that the poem is not only about the cultural change in Venice, but
about that at the heart of our own language and culture also. Again
the evolution of the mind of Europe is under scrutiny.

There is another side to this use of allusion. Those lines

> Defunctive music under sea
> > Passed seaward with the passing bell
> Slowly:

refer to a stage direction, not to life. As *Antony and Cleopatra* has
it, 'Music of the hautboys is under the stage'. Similarly, the reference
to the barge really refers to an account in a play of a scene which
we do not and in some ways could not witness, a scene which may
never have taken place in those terms. Eliot was well aware it was
all a business of transmission and reinterpretation of past interpretations as he shows in writing that 'Shakespeare acquired more
essential history from Plutarch than most men could from the
whole British Museum.'[24] Reading this poem urges us to ask what
'essential history' is and wonder if it has some ritual pattern. As
'literary allusions' the Shakespearian references point us back to
previous dramatic interpretations of events rather than directly to
'real events'. These are always lost under interpretation. We are
caught in a series of mimeses which push Burbank into the focus
of ritual, his actions sliding imperceptibly into reactions to and re-
enactments of a past rite whose origins we can never witness. A
movement downwards towards origins is, though, perceptible in the
poem. Burbank's 'descending' leads to 'fell' and passes 'under sea'

---

[24] *SE*, p. 17.

to a divine explanation of events and then to the mythological splendour of the ascent of the horses of the dawn (a little sabotaged by the questionable diction of 'Beat up'). But such an explanation is surely ridiculous to the modern reader forced to realize how far he stands from an age when mythological explanations were permissible. He is tempted to reinterpret the past in a way that makes it fit in more readily with the rendezvous at the small hotel. A parallel descent to first causes in the latter half of the poem locates a different underlying prime mover:

> On the Rialto once.
> The rats are underneath the piles.
> The Jew is underneath the lot.

This leads to the rather less splendid ascent of Princess Volupine 'To climb the waterstair' and to entertain

> Sir Ferdinand
>
> Klein.

the sound of whose surname and its positioning at the start of the final stanza aurally and visually rhymes with 'Declines' which similarly ends a sentence as the first word of stanza six. Climbing the waterstair, Princess Volupine moves on to the piles to meet the Jew who 'is underneath the lot'. Ascent and descent are alarmingly confused. Fittingly, at the poem's centre is the evolutionary joke,

> A lustreless protrusive eye
> Stares from the protozoic slime
> At a perspective of Canaletto.

The way the poem is ordered allows us momentarily to read it as if the juxtaposition was just that of eye and Canaletto, for the eye appears at first as an independent entity. But when we realize that the eye is most probably Bleistein's, we realize that it represents a degeneration from the expected human eye. The eye which seems able physically to perceive as it 'stares' but unable to appreciate what it sees links the modern hybrid, a world of complex cross-breeding inhabited by 'Chicago Semite Viennese', with the most primitive, 'protozoic' level of life. Again we are faced with a shift in cultural values. The Canaletto presumably has not altered, and men still come to look at it just as they repeat Shakespeare's words. Ritual observances continue awkwardly as in the clumsy Hakagawa-like

genuflexion, 'A saggy bending of the knees', but the original sense
of the ritual, as in so many of the cases described by Harrison and
Frazer, seems lost. This poem hints that we *can* go beyond the form
of the ritual only·to be horrified by the realization that there is nothing
there, or there is only some petty squalor, meaning having receded
out of sight if it was ever present in the first place. Evolution has
given us the Canaletto whose complex city perspectives recede
infinitely away and now it is itself seen regressing as

> The smoky candle end of time
> Declines.

Falling as decadent decline, degeneration, and as the sense of the
result of Original Sin is ubiquitous. 'They were together, and he fell.'
This is again a borrowing—from Tennyson, another staple of a
cultural tradition.[25] The realization of that fact is awkward since
Eliot's virtual repetition of Tennyson's line (which in 'The Sisters',
reads 'They were together, and she fell') in another context constitutes
an act of reinterpretation. Again the notions of legal fictions and
of rituals whose meaning may alter despite continuity of formal
behaviour are useful. Tennyson's poem, though, is about a woman
who lures a man to his doom, a man apparently of another race,
so that this allusion in turn provides a further interpretation of the
fates of Antony and Burbank. It may flesh-out Burbank's story, and
it may strip away some of the glory from the Shakespearian version
of Antony's. It does seem to strengthen the ritualistic patterning of
history. Life as ritual and art is a belief we associate very much with
those 'long-forgotten 'Nineties when sins were still scarlet' against
which Eliot professed to be reacting in 1917 when he declared that
their 'aesthetic eccentricities may now be ignored'.[26] But it seems
that his interest in primitive ritual had led him to place his own stress
on life as a ritual. For Eliot, though, this concern with ritual was
to have the sanction of the anthropologists' 'collective representation',
being generated like a *dromenon* by a group, a chorus across history.
Not Eliot alone, but Eliot, Shakespeare, and Tennyson produce the
poem. The combination seems to point to some underlying form of
'essential history' of which each individual provides his variant but
which can only be hinted at, not revealed, because when the voices

[25] Grover Smith (1974), p. 52.
[26] 'The Borderline of Prose', *New Statesman*, 19 May 1917, pp. 157–8.

join across time they never quite marry, though their coming together
is an attempt to generate something which like a collective emotion
'is necessarily felt as something more than the experience of the
individual, as something dominant and external'.[27] It might seem
that the primeval Fall underlies this poem's events as surely as Venice
underlies its epigraph, standing as the reality which underlies various
writers' interpretations. Yet even the meaning of that city seems
doubtful. Its bridges and canals are intact, but its meaning may have
altered. The final question,

> Who clipped the lion's wings
> And flea'd his rump and pared his claws?
> Thought Burbank, meditating on
> Time's ruins, and the seven laws

assumes a glorious past. Following the poem's ritualistic treatment
of history and treating it in accord with Eliot's method for ritual's
interpretation, if we take 'enough cross-sections to interpret a process',
and the poem invites us to do so, then we must be aware of a general
downward movement. The falling of Burbank, taking us down the
moral ladder, and the 'saggy bending of the knees' of Bleistein, taking
us down the evolutionary ladder, lead to the declining 'smoky candle
end of time' which prepares Burbank and the reader to ponder over
'Time's ruins', the etymology of 'ruins' being important. Royce's
seminars had acquainted Eliot with the second law of thermo-
dynamics if he had not known it before.[28] All time in this poem
seems to be a running-down. But the way in which present and
past mutually interpreting falls are linked together suggests that the
idea of the Fall itself may be illusory, a lost first cause. Burbank
falls; Hercules's departure marks Antony's decline. We see only
what is falling, never any Edenic first position fallen from. A glance
over the shoulder abolishes Eden. For though Burbank's fall, that
mere matter of crossing 'a *little* bridge' to a '*small* hotel' (my italics)
is a shrunken affair, lacking the panoramic mythological grandeur
of the Shakespearian Antony's, it none the less offers a stripped-
down representation of that event. We are invited to read back
destructively. If the previous state was really as fallen as this, is
the whole idea of a Fall illusory? If the unfallen state is being
pushed further and further out of sight, it may eventually vanish

[27] Harrison, p. 45.   [28] Grover Smith (1963), p. 150.

into unreality. If the continuing ritual represents a progressive pruning away of the trappings of mythology, then it would follow that no one has clipped the lion's wings or fleaed his rump or pared his claws since all these things, like classical-cum-Shakespearian mythology, are the ritual entrappings of the culture from which our own is descended, preserved among us as 'survivals'. But while externally unchanged they have ceased to be our central 'collective representations'. So that in a sense we have all been clipping, fleaing, and paring. The poem is funny, but at everyone's expense.

The effect of Eliot's reading about primitive religion is clear in 'Mr. Eliot's Sunday Morning Service' where, as in his Wundt reviews, he is concerned to stress the importance of the sexual element in religion. Coming from the reading of Frazer, Durkheim, and others, he does this by linking Christian rituals to sex and the continuance of vegetation.

> Along the garden-wall the bees
> With hairy bellies pass between
> The staminate and pistillate,
> Blest office of the epicene.

From the opening word, 'Polyphiloprogenitive', the stress on religion as sexually oriented is evident. There is a conflict of languages within the poem. The overarticulate 'Superfetation' and the physically direct 'hairy bellies' both point in the direction of sex. Formal games too consciously played increase the sense of unease. Language flaunts itself, embroidering ritual in alliterative use of *s* and *p* through stanzas five and six. But the linguistic oddness is deliberate. Language itself, word following Word, is giving birth to new forms — 'piaculative' is listed in the *OED* as 'rare', this poem being its only cited occurrence. Most likely it derives from Durkheim whose penultimate chapter in *The Elementary Forms of the Religious Life* is entitled 'Piacular Rites and the Ambiguity of the Notion of Sacredness'.[29] The new word achieves mutant birth in a poem about fertility and birth, stressing the sexual element in religion. For Durkheim part of the ambiguity of the sacredness involved in the aboriginal fertility rites was that

The same results are obtained by fasts, abstinences and self-mutilations as by communions, oblations and commemorations . . . However complex

---

[29] Durkheim (1971), p. 389.

the outward manifestations of the religious life may be, at bottom it is one and simple. It responds everywhere to one and the same need, and is everywhere derived from one and the same mental state. In all its forms, its object is to raise man above himself and to make him lead a life superior to that which he would lead, if he followed only his own individual whims: beliefs express this life in representations; rites organize it and regulate its working.[30]

This passage sheds much light on the method of 'Mr. Eliot's Sunday Morning Service' where the self-mutilation of 'enervate Origen' is placed in the same lineage as the sexual origin of 'the Word', and where the initial 'sapient sutlers of the Lord' who 'Drift across the window-panes' is a passage with an ambiguous, or better ambivalent, reference, since it holds together in one term both the 'sable presbyters' who bring offerings to church and 'the bees' who bring pollen from one part of the plant to another and so perform the 'Blest office of the epicene'. Religion is presented as a response 'everywhere to one and the same need': the need for rituals of fertility. Durkheim commented on the difficulty of regarding 'an enlightened Catholic of to-day as a sort of retarded savage', but he provided the reminder that 'the most primitive religions are not the only ones which have attributed this power of propagation to the sacred character'.[31]

As in 'Burbank' mythology was presented as painting, so here we have 'Designed . . . The nimbus of the Baptized God'—perhaps based on 'The Baptism of Christ' by Piero della Francesca whose work hung in the National Gallery's Umbrian School room marked in Eliot's *Baedeker*.[32] The word 'nimbus' meaning both halo and rain-cloud is carefully deployed. The primitive ceremony which came soonest to Eliot's mind, whether in his 'Beating of a Drum' or in 1926 when he spoke of savages who 'believe that the ritual is performed in order to induce a fall of rain', was rain-making, probably because he had read about it for his 1913 seminar paper. Frazer's *Dying God* describes rituals where drums are beaten to beg rain from heaven. *Themis* describes various rain-making rites, including a Christianized one which involves entreating 'S. John Baptist to baptize the Holy Child'. Lévy-Bruhl links rain-making with Christian practice; writing of the aboriginal *intichiuma* ceremonies he points out that 'Nothing

---

[30] Ibid., p. 414.   [31] Ibid., p. 322.
[32] Suggested in Stephen Spender, *Eliot* (Glasgow: Collins, 1975), p. 57; see Eliot's *Baedeker* (n. 17 above), p. 172.

is more widespread than practices having as their object the cessation of drought, and the assurance of rain: (we see this even yet in our own Rogations) . . . '[33] Durkheim, following Spencer and Gillen, described the aboriginal 'Intichiuma of rain'.[34] 'The object of the whole rite is to represent the formation and ascension of clouds, the bringers of rain.'[35] The whole direction and tone of Durkheim's investigation was 'to find a means of discerning the ever-present causes upon which the most essential forms of religious thought and practice depend'. In a poem continually relating Christian religion to fertility ritual Eliot's painted annunciation hints at the most primitive rites not only in its 'nimbus' but also in its drought-stricken landscape, where 'The wilderness is cracked and browned.'

The coming of God is linked with the making of rain both through the 'nimbus' and in what follows.

> But through the water pale and thin
> Still shine the unoffending feet
> And there above the painter set
> The Father and the Paraclete.

This presentation allows the hint of primitive ritual to remain in the background, just as later Eliot was to imply with deliberate provocation that behind the High Mass lay, if one went back far enough, 'the Australian ceremonies described by Spencer and Gillen and Hewett [*sic*]'.[36] This poem foreshadows the method of the later, greater poetry of *The Waste Land* in trying to hold the most primitive and most developed in one by letting the former show through the latter and the development of one out of the other be seen. Just how strongly Eliot was spurred on to emphasize the connections between sex and religion at this time is seen in 'The Hippopotamus' (1917) which contains one of his most shocking jokes:

> At mating time the hippo's voice
> Betrays inflexions hoarse and odd,
> But every week we hear rejoice
> The Church, at being one with God.

Throughout these quatrain poems sex and religion mix, so that always 'The nightingales are singing near / The Convent of the

---

[33] Lévy-Bruhl, p. 250.   [34] Durkheim (1971), p. 353.
[35] Ibid., p. 354.   [36] 'The Ballet', *Criterion*, Apr. 1925, p. 441.

Sacred Heart.' Whether there is anything beyond the sexual is uncertain. Christ's feet in 'Mr. Eliot's Sunday Morning Service' are 'unoffending' which may mean without sin, but might also mean just harmless or inoffensive, devoid of power in a way paralleling Origen's sexual impotence; the apparent presence of God need not guarantee Christ's power. Jane Harrison had seen the Olympians as moving from fertility *daimones* to *objets d'art*.[37] In Eliot's poem the status of God as *objet d'art* is stressed ('And there above the painter set / The Father and the Paraclete') so that for all that we may penetrate beyond that either to gesso ground or vestiges of primitive fertility ceremony, the ultimate point of origin, the postulated God, is to be explained simply in terms of sexuality, or else remains unreachable and inexplicable. Eliot, who was interested in *mana* and *orenda*, worried that Durkheim's theory of totemism in terms of religious force might prove only 'an admission of the inexplicable'.[38] The learned gloss, 'Superfetation of τὸ ἕν' is deliberately awkward. It shows how the Word could give birth to commentators' words, as could the castrated Origen, but gives no acceptable explanation of how it could produce Origen. The poem enacts various creations as parts of an apparently continuing ritual pattern; it is polyphilo-progenitive in detailing the creation of the Word, Origen, the painting, and flowers; we seem able to trace all back to their cause, yet ultimately the cause is only a formal utterance, 'In the beginning was the Word', whose repetition takes us no further. We are back with the problem which Eliot had considered in March 1914 when one of his seminar colleagues, Sen Gupta, had read a paper that Eliot had written dealing with the work of Lévy-Bruhl 'on primitive race-psychology' and the 'law of participation', a paper in which it was stated that, 'Causality is something which can be explained away but not explained. Perhaps it is due to superstition, but can we do without such superstitions?'[39] The question of whether or not 'In the beginning . . . ' is superstition is forced on us by the poem but left unanswered though it appears that to go back beyond Origen would be immensely difficult.

For Eliot in this poem the only way to banish endless debate lies not in accepting dogma, but in fleeing from speculation to Sweeney

[37] Harrison, table of contents, p. xxviii.
[38] For Eliot's interest in *mana* and *orenda* see e.g., IPR, p. 13, and 'Durkheim'.
[39] Grover Smith (1963), p. 138.

in his bath. With the comic appearance of this modern barbarian
the resolutely physical takes over; we move back to the level of the
naked man with no interest in the 'polymath'. Yet Sweeney, found
in the essential element of fertility for both aborigine and Christian,
is also a mover of waters. Ironically, though Sweeney appears to
promise a way out, he may only lead the sensitive reader back into
the circle of sexual-religious speculations.

Neither the religious nor the sexual anxiety in Eliot's poems
came simply from his reading, but that reading allowed him to give
supra-individual expression to his personal pain; he clothed his own
cry in the language and mythology of great traditions and their
interpretations, making it 'impersonal'. Eliot was doing this before
he had articulated a theory about it. His reading also helped him
articulate his theory. Durkheim wrote that, 'there is something
impersonal in us because there is something social in all of us, and
since social life embraces at once both representations and practices,
this impersonality naturally extends to ideas as well as to acts.'[40]
It was as a possession of society that mythology and the great art
of the past was of use to the poet, since as such its strength and
currency was established and purged of personal intrusions. Eliot
wrote in 1916 that mythology was 'dangerous literary material' and
had to be either a mythology in which the writer believed or else
one in which a people had once believed. 'A mythology cannot be
created for literary purposes out of whole cloth; it must be the
work of a race.'[41] Admiring Durkheim's stress on the need for
community, Eliot wrote that, 'For the savage or the civilized man,
a solely individual existence would be intolerable: he feels the need
of recreating and sustaining his strength by periodic refuge in another
consciousness which is supra-individual.'[42] So the poet recreates his
strength in the tradition, which can provide a refuge and a strength,
yet which can also appear founded on emptiness. In voicing his
theory of impersonality, and in searching for terms of extreme yet
impersonal expression anthropology was important in terms of the
ideas and structures which it provided, but also in terms of its
vocabulary again often used for the communication of deep distress.
So Eliot, in 'Ode', a poem dealing with a tortured wedding night,

[40] Durkheim (1971), p. 446.
[41] 'Mr. Doughty's Epic', *Manchester Guardian*, 23 June 1916, p. 3.
[42] 'Durkheim'.

has a protagonist apparently lacking in all inspiration except that coming from the 'bubbling' of a river described as 'mephitic'.[43]

In *Adonis Attis Osiris*, Frazer saw the goddess Mefitis as personifying 'mephitic vapours' and described her temple 'where the exhalations . . . were of so deadly a character that all who set foot on the spot died'.[44] Mefitis's lethal pool 'continually bubbles up with an explosion like distant thunder'. But it could also serve for the purposes of inspiration. The sick left there 'sometimes . . . were favoured with revelations in dreams . . . To all but the sick the place was unapproachable and fatal.'[45] Frazer also mentions a sanctuary whose mephitic area contains only 'the eunuch priests of the Great Mother Goddess', who have 'a look on their faces as if they were being choked'.[46] Sleeping with the succuba who may be disembowelled but also disembowelling, Eliot's speaker is in a position analogous to that of the eunuch priest; poetry like sex here points to horrible death and finds expression for that concept in the language of the rituals of primitive religion and mythology. Except that the meaning of these has waned. The rituals are repeated, the mephitic river risked, but only to find it finally 'uninspired'.[47]

Anthropology both masked and revealed Eliot's own concerns. It was a lens through which he could view life, literature, and history, often with mischievous irony. In view of all his interest in connections between worship of totemic plants, rain ceremonies, and modern life, it becomes clear why 'Gerontion' is sparked off by a passage from Benson's *Fitzgerald* where Benson is quoting Fitzgerald.

'Don't you love the Oleander? So clean in its leaves and stem, as so beautiful in its flower; loving to stand in water, which it drinks up so fast. I rather worship mine.'

Here he sits, in a dry month, old and blind, being read to by a country boy, longing for rain . . .[48]

Fitzgerald's oleander-worship and longing for rain are like those 'survivals' Eliot discussed in 1917. Fitzgerald's brain was also haunted by survivals of past literature, 'The old music of bygone singers, rich haunting sentences of old leisurely authors, rang in his brain, and came unbidden to his pen.'[49] 'Gerontion' presents a little old man,

[43] 'Ode', *Ara Vos Prec* (London: Ovid Press, 1919).
[44] Frazer, *Adonis Attis Osiris*, i. 204.   [45] Ibid., pp. 204–6.
[46] Ibid., p. 206.   [47] 'Ode'.
[48] A. C. Benson, *Edward Fitzgerald* (London: Macmillan, 1905), p. 142.
[49] Ibid.

whose speech is riddled with other men's words, waiting 'in a dry
season' for rain. We appear to be in the cosmopolitan city world
of the uprooted.

> My house is a decayed house,
> And the Jew squats on the window sill, the owner,
> Spawned in some estaminet of Antwerp,
> Blistered in Brussels, patched and peeled in London.

Mr Silvero, Hakagawa, and the other figures stress this up-rootedness.
Yet, ironically, 'waiting for rain', we are rooted in the most primitive
situation. The communion ceremony is seen in terms stressing the
physicality of Frazer's 'Eating the God'.[50] Rendel Harris's gods were
originally plants. Discussing the sacrificial meal, Durkheim, Harrison,
and others saw eating totemic animals or plants as a primitive
communion, as Eliot knew. Eliot noted that in Australian religion
Durkheim found the essential elements of all religion, and that for
Durkheim *communion*, not worship, was the essential sentiment.[51]
Eliot's 'word, / Swaddled', with its suggestion of Christ's 'swaddling
clothes' (Luke 2: 12) places us firmly within the area of fertility and
birth, but Christ in 'Gerontion' comes not as baby but as frightening
beast, associated, like totemic animals, with fertility. Reviving
vegetation is emphasized, being bound up in 'judas' with the Christian
(and New England)[52] story. But the gap between Christ and the
eating not only emphasizes the vegetation, but makes it seem for a
moment that it is the plants which are devoured:

> In the juvescence of the year
> Came Christ the tiger
> In depraved May, dogwood and chestnut, flowering judas,
> To be eaten, to be divided, to be drunk

Gerontion, in the barren wilderness of modern life, the world of
Antwerp, Brussels, and London, is terrified of the cycle of fertility
which also traps him. For him 'The tiger springs in the new year.'
His years are not modern calendar years, but governed by the rebirth

[50] Sir J. G. Frazer, *The Golden Bough, Third Edition, Part V, Spirits of the Corn
and of the Wild* (London: Macmillan, 1912; 2 vols.) ii. 48–108.
[51] 'Durkheim'; see Durkheim (1971), pp. 337–41, and cf. King, *passim*, esp.
pp. 149 ff.
[52] The plants in 'Gerontion' are from Adams's *Autobiography*: see Grover Smith
(1974), p. 62.

of vegetation—new year comes in 'depraved May'. Rebirth for him would be fatal, in the word of a later poem 'like Death, our death'.[53] Both Christ and the fertility cycle seem terrifying. Yet when he seeks refuge in history, other men's deeds and words, Gerontion finds nothing with which he can connect, only vacancy, vanity, and inane deception: the emptiness of the uninspiring, fragment-bearing wind that blows through much of Eliot's poetry.

The shrunken personal emptiness in an arid landscape, the hungering after other men's experience, and the combined fear and fascination which Gerontion feels for threatening rebirth seem to grow, partly at least, from Eliot's own situation. In London he found himself chronically overworked, unhappily married, and a failure in his parents' eyes. 'Gerontion' looks much towards death; Eliot's father died in January 1919. To him Eliot dedicated *The Sacred Wood*. Unable even to enlist because of ill health, Eliot had found work as a bank clerk while others fought in the warm rain. He found the bank's routine congenial, keeping greater worries at bay. Concentrating attention on tasks before him avoided the resuscitation of more painful parts of his life. Such resuscitations, though, were inescapable. Reviewing Henry Adams's *Autobiography* in 1919 he attacked the Bostonian world and, covertly, himself, as he savaged Boston Unitarianism along with Adams who 'abandoned lecturing at Harvard', and whose researches into primitive mythology 'turned to ashes in his mouth'. Doubt choked Adams. Like Gerontion, 'there is nothing to indicate that [his] senses either flowered or fruited.'[54] In 1919 Eliot felt himself to be in a similar position.

This seems ironic, but letters and later statements show how little faith Eliot had in his own literary output, however confident he might sound in verse, or in the *ex cathedra* pronouncements of 'Tradition and the Individual Talent'. In 1922 he threatened to give up literature.[55] His fascination with religious ideas brought no peace. He was confronted with the failure of his personal life (impersonalized in 'Ode' and elsewhere) and, it seemed, of his public life. His reading of Henry Adams was not the only painful revival of memory or confrontation with his own apparent failure which he had to endure in 1919.

[53] *CPP*, p. 104.
[54] 'A Sceptical Patrician', *Athenaeum*, 23 May 1919, pp. 361–2.
[55] Letter to Sidney Schiff, 20 Apr. 1922 (British Library).

124 *Waiting for Rain*

In June of that year Eliot met a figure from what must now have seemed to the London banker a remote part of his life. At Harvard, another friend recalled, Eliot's

really closest friend was Harold Peters, and they were an odd pair. It was Peters who chided him about his frail physique, which led to his regular attendance at August's Gymnasium, which was in the basement of Apley Hall. He took this work seriously and developed into quite a muscular specimen. It also led to some boxing lessons somewhere in Boston's South End. He took up rowing in a wherry, and finally worked up to a single shell. Peters also introduced him to small-boat cruising and they made many cruises between Marblehead and the Canadian border. On one of these trips, in a 19-foot knockabout, before the days of power, they rounded Mt. Desert Rock in a dungeon of fog, a rough sea and a two-reef breeze.[56]

Such voyages would reappear in Eliot's poetry, and not only in the hunt for lost childhood and youth that fills 'Marina'. In late June 1919 Eliot went to Garsington, but cut short his visit to stay with Peters who was about to be demobbed from the US Navy.[57] Vivienne found Peters extremely boring, but the next day, Sunday, Peters and Eliot, who was looking very ill, went off to Greenwich, a choice of venue which suggests that the two men talked about sailing for much of the time. Certainly they would have had much to talk about, recalling voyages up the New England coast; how they had dared storms together and mixed with tough Maine characters at the Jonesport summer ball, or how one of Peters's cruises had been christened after the four B's which constituted their provisions: beans, bacon, bread, and bananas.[58] Peters would have had much else to tell Eliot. In 1915 he had sailed to South America on one of the last square-riggers. In the war he had sailed to Cuba and between the east coast of America and Scotland. Now he was about to return to America, a bachelor who would sail round the world on various cutters and schooners. The next day, Monday 23 June, Peters left for America leaving Eliot an unhappily married London bank clerk on the fringes of a circle which he described as a 'Bloomsburial'.[59]

---

[56] T. S. Eliot, R. W. H. and L. M. L[ittle], tribute (C677b), p. 53.
[57] Information from Vivienne Eliot's 1919 diary (Bodleian) and Anon., obituary for 'Harold Peters', *Harvard College Class of 1910: Thirty-fifth Anniversary Report* (Cambridge, Mass: Cosmos Press, 1946), p. 183.
[58] Correspondence with L. M. Little (Houghton).
[59] See 'Harold Peters' and Vivienne Eliot's diary (n. 57) above); 'Shorter Notices', review of Clive Bell, *Potboilers*, *Egoist*, June/July 1918, p. 87.

Eliot's meeting with Peters brought a painful revival of past good times. By 1 June 'Gerontion' had been half finished.[60] It seems to have been completed by the start of July.[61] In the last stages of composition, after the destruction of the 'society' names, 'De Bailhache, Fresca, Mrs. Cammel', Eliot amended his original version, adding some lines which surely came as a result of his meeting with Peters:

> . . . Gull against the wind, in the windy straits
> Of Belle Isle, or running on the Horn.
> White feathers in the snow, the Gulf claims,
> And an old man driven by the Trades
> To a sleepy corner.[62]

In late 1919 Eliot resolved to begin a long poem.[63] His meeting with Peters also seems to have sparked off the long voyaging section of 'Death by Water' in the *Waste Land* manuscripts, which would be united with the fate of the ancient Phoenician sailor, Phlebas, and details of which would find their way into 'Marina' and 'The Dry Salvages'. This section, describing in accurate seamen's language a voyage past The Dry Salvages in a fishing boat loaded with 'canned baked beans' combined Eliot's own sea knowledge with that of the Gloucester dory fishermen, and bound these up with the 'well-told seaman's yarn' of Dante's Ulysses and with Tennyson's Ulysses (who may be related to Gerontion).[64] The eventual dropping of this passage, retaining only the Phlebas section, makes Part IV of *The Waste Land* entirely impersonal, but underlying it was a resuscitation of personal memories.

Further memories were revived by a visit from his mother in the summer of 1921 which brought 'another anxiety as well as a joy' when Eliot was working on what would become *The Waste Land*.[65] Eliot's 1921 meeting with Dr Roger Vittoz in Lausanne must have once more forced him to confront his own history and ambitions. He also found peace there and renewed an old pleasure, by boating at Lugano where he met Hesse.[66] In Switzerland, though he met 'people of many nationalities, which I always like', Eliot felt

---

[60] See Gordon, p. 100 n. 31.
[61] Eliot sent a version around this time to Schiff; see his letters to Schiff, 16 July 1919 and 25 July 1919 (British Library).
[62] These lines appear in only one of Berg MS versions; they have been added.
[63] *Facsimile*, p. xviii.   [64] Ibid., pp. 54–69, 128.   [65] Ibid., p. xx.
[66] Letter to Schiff (n.d.) (British Library).

that he had released Vivienne from the strain of being with him.[67]
He too had time to think and to write.

In the *Waste Land* which Eliot eventually produced, the pain of
personal resuscitations persists, but that pain is bodied forth using
a structure of ideas that results from the coming together in Eliot's
mind of various views of the savage and city. We now know that
around 1922 Eliot was close to becoming a Buddhist.[68] His own
studies in Japanese Buddhism, his attending Oxford's Buddhist
Society, and his reviewing of Oriental books are indicators of such
interests. But they are present, too, in his literary theorizings.
'Tradition and the Individual Talent' is grounded on the idea that
'not only the best, but the most individual parts of his [an author's]
work may be those in which the dead poets, his ancestors, assert their
immortality most vigorously'.[69] In the same year, writing of how
as good writers 'we have not borrowed, we have been quickened,
and we become bearers of a tradition', Eliot complains (before
quoting a revoicing of Seneca by Chapman which would be used
in 'Gerontion') that in contemporary poetry, 'No dead voices speak
through the living voice; no reincarnation, no re-creation.'[70] Reincarnation and re-creation would be crucial to *The Waste Land* where
Buddhist ideas would have an important place and where various
voices of the dead, from that of Dante to that of Mayne Reid, would
play their part in a sort of literary metempsychosis, not entirely unlike
that used by Joyce in *Ulysses*, which Eliot was reading at the time.[71]
For Eliot, though, such ideas of tradition and reincarnation would
be bound up with heredity at the most basic evolutionary level and
with the continuing cycle of renewal and death which he found in
the pages of the anthropologists. These pages revealed rituals which,
Jessie Weston informed him, had survived the passage *From Ritual
to Romance*, remaining hidden in modern literary forms in the world
of Tennyson and Wagner.[72]

When Eliot became a Christian in 1927 he declared that he found
in reading Paul Elmer More, with whose *Shelburne Essays* he had

[67] *Facsimile*, p. xxii.
[68] 'The Modern Dilemma' (C348), as summarized by Caroline Behr, *T. S. Eliot:
A Chronology of his Life and Works* (London: Macmillan, 1983), p. 44: cp. Stephen
Spender, *Eliot* (Glasgow: Collins, 1975), p. 26.
[69] *SE*, p. 14.   [70] 'Reflections' (C84), p. 39.   [71] *Facsimile*, p. xx.
[72] Jessie L. Weston, *From Ritual to Romance* (Cambridge: Cambridge University
Press, 1920), p. 188.

shown familiarity in 1916, the work of someone who had travelled by almost the same route, to virtually the same conclusions.[73] From More Eliot seems to have taken the word 'anfractuous', and read that the poet undergoes 'a partial dissolution of his own personality'.[74] Like Eliot, More had been interested not only in the classics and Christianity, but also in the East. The More who was interested in a religious 'bridge between the Orient and the Occident', and who juxtaposed Augustine and Buddha paralleled the Eliot of *The Waste Land*.[75] But More also linked eastern thought about reincarnation to Darwinism, since both stressed the presence of the past and the idea that the most primitive forms persisted in the modern. More feels this presence of the past when 'we look into the eyes of love' and obtain 'for one supercelestial moment—the glimpse of a reality never before imagined, and never again to be revealed'.[76] He was fascinated by Lafcadio Hearn's having

brought together into indissoluble union our Western theory of Darwin and that strange doctrine of metempsychosis which was carried to Japan with Buddhism . . . To understand the tremendous realism of horror and gloom connected with this doctrine of everlasting birth and death, and re-birth, one must go to the burning valley of the Ganges . . .[77]

In *The Waste Land* Eliot would go there, armed with anthropological ideas gained from Frazer and others. More wondered if Hearn had

introduced a new element of psychology into literature? We are indeed living in the past, we who foolishly cry out that the past is dead . . . Mr. Hearn shows how even the very beasts whom we despise as unreasoning and unremembering are filled with an inarticulate sense of this dark backward and abysm of time . . . By reason of this terror the savage trembled before the magician who seemed to have penetrated the mysteries of nature about him.[78]

For Eliot, Frazer's work was of an importance paralleling the psychology of 'the complimentary [*sic*] work of Freud—throwing

[73] 'Paul Elmer More', *Princeton Alumni Weekly*, 5 Feb. 1937, p. 373.

[74] Paul Elmer More, 'The Scotch Novels and Scotch History', *Shelburne Essays, Third Series* (New York: G. P. Putnam's Sons, 1905) pp. 83 and 85.

[75] Paul Elmer More, 'St. Augustine', *Shelburne Essays, Sixth Series (Studies of Religious Dualism)* (New York: G. P. Putnam's Sons, 1909), p. 73.

[76] Paul Elmer More, 'Lafcadio Hearn', *Shelburne Essays, Second Series* (New York: G. P. Putnam's Sons, 1905), p. 63.

[77] Ibid., p. 71.   [78] Ibid., pp. 65–6.

its light on the obscurities of the soul from a different angle'. Frazer 'has extended the consciousness of the human mind into as dark a backward and abysm of time as has yet been explored'.[79]

Ideas from Buddhism, evolutionary theory, and anthropology combined in *The Waste Land*. I have already remarked on the poem's opening where Mendel and Rendel Harris combine to reincarnate past voices, plant breeding and anthropology shedding a new light on the dead — from the Chaucer of the 'Prologue' to James Thomson and the Fitzgerald whose stanza More had quoted when writing about the growth of trees and flowers in an essay of 'Saint Augustine':

> Now the New Year reviving Old Desires,
> The thoughtful Soul to Solitude retires
>   Where the White Hand of Moses on the Bough
> Puts out, and Jesus from the Ground suspires.[80]

Eliot's Bough was *The Golden Bough* which had plenty to say about the rebirth of gods accompanying vegetation. In the final version of the poem Buddha meets St Augustine, while Christ and Buddha merge in the 'I' of the abandoned 'I am the Resurrection and the Life' fragment.[81] The anthropological elements in the final poem are also present in the collection of manuscripts from which it emerged. We see there how Eliot indulged his sense of irony by resuscitating the voices of older poets through his anthropological interests and satirizing 'a time, barren of myths' and a world of those 'illimitable suburbs' which he saw in 1921 as filled with the public school's 'petrified product'.[82]

'Exequy' uses the essentially Frazerian structure of 'a man-god'.[83] Its 'sacred grove' is a phrase straight from *The Golden Bough*. Frazer's evocative and often imaginative description of Nemi's grove under 'the dreamy blue of Italian skies' ensures that Eliot's poem takes place in 'that Italian air'.[84] Yet we are uncertain of a temporal context for the poem. If we are dealing with a 'suburban tomb', then the word suburban suggests, though it need not mean, the life of the modern city. Later, the 'Austrian waltz' will also suggest this. As the anthropologists whom he had read had used poetry to exemplify

---

[79] 'A Prediction' (C153), p. 29.
[80] Fitzgerald, quoted in Paul Elmer More, 'St. Augustine', p. 78.
[81] *Facsimile*, pp. 110–11.
[82] 'Notes on Current Letters', *Tyro* [Spring 1921], p. 4.
[83] *Facsimile*, pp. 100–3.   [84] Frazer, *The Magic Art*, i. 9.

various rituals, so Eliot does the same with an ironic bite. In the
second stanza we find him still in the classical world, though this
time it is the Greek rather than the Italian, and it begins by being
the Greek seen not through the eyes of Frazer, but through the eyes
of Keats. Eliot's second stanza opens:

> When my athletic marble form
> Forever lithe, forever young,
> With grateful garlands shall be hung . . .

There are quite sufficient echoes here to direct the reader to Keats's
'Ode on a Grecian Urn' where he finds the urn 'With brede / Of
marble men and maidens overwrought, / With forest branches . . . '
and finds too scenes of love 'For ever warm and still to be enjoyed,
/ For ever panting, and for ever young—' as well as the heifer
approaching the altar 'with garlands dressed'. Eliot, though, is
determined, like the anthropological writers he had been reading,
to make plain the root of the custom, which he does in the next line,
'And flowers of deflowered maids'. As in the Frazer of 'The Influence
of the Sexes on Vegetation', the strong links between sex and religion
are revealed.[85] This is also in line with Eliot's own criticism of the
lack of such relationships in Wundt's work. But if Eliot in the poem
has adopted the personality of a fertility god, this god is a peculiarly
Prufrockian one in the sense that for all 'The constant flame shall
keep me warm,' he remains not simply a minor divinity, neither being
nor meant to be Prince Hamlet, but also, for all the lovers' attentions,
an impotent ghost, 'A bloodless shade among the shades / Doing
no good, but not much harm'. The poem is an undermining of
traditional notions of love, through repeating love's rites in such a
way as to show them as meaningless. The Passionate Shepherd,
Marlowe, invoked his pastoral love to be with him 'By shallow rivers,
to whose falls / Melodious birds sing madrigals'. In Eliot's poem
it is 'While the melodious fountain falls' that love is made, but we
are forced to be conscious of an artifice '(Carved by the cunning
Bolognese)' which suggests that the apparently primitive fertility ritual
where 'The Adepts twine beneath the trees / The sacrificial exercise'
has become a decadent pleasure, rather than a genuine ritual. It is
ritual decayed into spectacle where not continuity but surprise has

---

[85] Ibid., ii (title of ch. XI).

become the ultimate consideration, tradition and originality having their horns uncomfortably locked in the phrase 'invariable surprise':

> They terminate the festivals
> With some invariable surprise
> Of fireworks, or an Austrian waltz.

In 1921, *The Waste Land* clearly on his mind, Eliot complained that the anti-Georgian poets were 'mostly such as could imagine the Last Judgment only as a lavish display of Bengal lights, Roman candles, catherine-wheels, and inflammable fire-balloons. *Vous, hypocrite lecteur . . .* '[86]

The last stanza of 'Exequy' is a summing-up of the kind of rituals described by Frazer in *Adonis Attis Osiris* where in the declining year the representative of the year-spirit was put to death, often by burning. This ritual suicide is seen as something deeper than the sexual rituals described in earlier stanzas, yet in Frazerian terms, it is itself a ritual performed in connection with the revival of vegetation in spring. The visitor of the last stanza comes 'more violent, more profound, / One soul, disdainful or disdained,' and in the condition of the year-spirit or ἐνιαυτὸς δαίμων, 'his shadowed beauty stained / The colour of the withered year', to go to a death which places him in the position of savage sacrifice and, for he is surely related to the saints of Eliot's other early poems, martyr 'Self-immolating on the Mound'. This visitor is himself unable to escape from the sexuality found to permeate even the most intense religious belief. Eliot was, of course, familiar with the association of death and sexual climax, having played with the idea in that murder of 'Nocturne' which calls forth the explanation

> 'The perfect climax all true lovers seek!'

but the nervous mark of this humorous poem's immature irony has been replaced by the much more horrific denial of the meaning of the action of the ritual

> Upon the crisis, he shall hear
> A breathless chuckle underground.
> ~~SOVEGNA VOS AL TEMPS DE MON DOLOR~~

In his draft Eliot has replaced this last cry with

---

[86] As n. 82 above.

~~Consiros *vei* la pasada folor.~~

—'in thought I see my past madness'; these two cries are from the Arnaut Daniel speech ending *Purgatorio* XXVI, where the poet condemned to the circle of lust prays for the hearer's consideration and looks forward to the release from the tortures meted out to the lustful. This passage haunted Eliot who was to quote from it in the original title of *Poems, 1920*, as well as in *Ash-Wednesday*.[87] As it is used in the draft poem it seems to imply that not only are the decadent versions of primitive sexual rituals inane, the god dead and impotent, however passionate the rituals of his worship, but that even the ultimate act of martyrdom is a sexual indulgence. Though the poetic richness of 'Exequy' would be surpassed in *The Waste Land*, a far subtler piece of writing, the emptiness of the lesser work and its pained despair would be repeated in the masterpiece. *The Waste Land* is not a Christian poem.

By the time when he began to work most intensively on what would become *The Waste Land*, Eliot was neither god nor martyr. Instead, he was living in a north London flat and working weekdays in a City bank, returning home at night to a difficult marriage. London itself, however, may have furnished encouragement for the lines Eliot's thought was taking. Visiting the Notting Hill Victorian mansion of Ford Madox Ford, Eliot saw not quite a sacred grove, but a patch of grass surmounted by Gaudier-Brzeska's 'Hieratic Head of Ezra Pound' which Wyndham Lewis described as 'Ezra in the form of a marble phallus'.[88] While in 1920–1 Eliot admired both Edward Wadsworth's drawings of industrial landscapes and Picasso's primitivism, he found himself living the life of Thomson or of the city clerks he had read about in Conan Doyle, Davidson, Conrad, and elsewhere.[89] Haunted by the voices of the past, Eliot had come to inhabit the landscape of his earlier reading. From that landscape, masking the real London around him, grew *The Waste Land*.

What appears to be one of the earliest sections of the manuscripts the section beginning 'So through the evening, through the violet air'

---

[87] *Ara Vos Prec*; *CPP*, p. 94.

[88] Wyndham Lewis, 'Early London Environment', in Tambimuttu and Richard March, p. 27.

[89] 'London Letter', March 1921, *Dial*, Apr. 1921, p. 453; letter to Sidney Schiff, 12 Jan. 1920 (British Library).

(dating from *c*.1914[90]) contains an arresting image which would find its way into the final poem.

> A woman drew her long black hair out tight
> And fiddled whisper music on those strings.

This image conflates two images from Kipling's short stories, images which Eliot would again juxtapose in 1941.

Compare the description of the agony in *In the Same Boat* (a story the end of which is truer to the experience than is the end of *The Brushwood Boy*): 'Suppose you were a violin string — vibrating — and someone put his finger on you' with the image of the 'banjo string drawn tight' for the breaking wave in *The Finest Story in the World*.[91]

'In the Same Boat' has Kipling's Conroy exclaim just before he meets the hysterical, drug-taking woman, Miss Henschil,

'I'm no musician, but suppose you were a violin-string — vibrating — and some one put his finger on you? As if a finger were put on the naked soul! Awful!'[92]

Eliot transfers this, yet another image of terrible pain, to an almost surreal context in his poem, but the woman who draws her long black hair out tight is related to that other hysterical woman of 'A Game of Chess', since she too is seen brushing her hair. Eliot's youthful familiarity with Kipling's stories was clear in his 1909 student essay on that author's prose.[93] Eliot was particularly interested in Kipling's use of the supernatural, 'one of his most effective tools', and saw him as roaming 'from Greenland to the South Seas in search of effects'. Stories which Eliot knew then such as 'They' and 'The End of the Passage', 'one of the most striking tales of fear that I have ever read', would haunt his poetry.[94] The younger Eliot tried to revolt against Kipling, as against the Nineties, but the older had a 'feeling of destiny' that he was linked to Kipling.[95] Certainly ' "The

---

[90] *Facsimile*, pp. 112–15; the dating on p. 130 is followed by Gordon, p. 143.
[91] *OPP*, p. 239.
[92] Rudyard Kipling, 'In the Same Boat', *The Bombay Edition of the Works of Rudyard Kipling; Vol. XXIV, A Diversity of Creatures* (London: Macmillan, 1917), p. 55. Grover Smith (1983), p. 73 draws attention to the Kipling image in 'The Finest Story in the World'.
[93] The unpublished MS of 'The Defects of Kipling' (Mar. 1909) is now in Houghton. Extracts appear in J. Donald Adams, *Copey of Harvard* (Boston: Houghton Mifflin Company, 1960), pp. 153–64.
[94] Ibid., p. 161.   [95] 'The Unfading Genius' (C632), p. 10.

Finest Story in the World"' with its hero who emerges as the unlikely reincarnation of a Norseman from Greenland and a galley slave was to have considerable repercussions.

The hero of Kipling's story is a London bank clerk who is also an ambitious poet. 'His name was Charlie Mears; he was the only son of his mother who was a widow, and he lived in the north of London, coming into the City every day to work in a bank.'[96] The widowed mother, the residence in north London, and the daily travel to work in the London bank were all part of Eliot's own experience at the time of the composition of most of *The Waste Land*. Charlie's 'Business took him over London Bridge.'[97] His poetic efforts are mocked by the narrator of the story, who is none the less fascinated by what stimulates Charlie's writing. Like Eliot, this narrator goes to the British Museum, and like Eliot he has an interest in metempsychosis. He introduces Charlie to 'Mortimer Collins's "Transmigration", and gave him a sketch of the plot before he opened the pages.'[98] Certainly Eliot's studies in Sanskrit and Pali at Harvard under Lanman and Woods had given him a thorough knowledge of Indian thought, but it is probable that Kipling's version of metempsychosis had at least an equally important effect.[99]

Kipling's story presents, through the consciousness of Charlie, extreme historical juxtapositions. Charlie Mears emerges as the reincarnation not only of a galley slave from ancient Egypt, but also of a seaman from the Vinland Sagas. These past lives are confused with the life of a modern bank clerk. So, just before speaking of the Skroelings (that is, Skrælings of *The Vinland Sagas*) Charlie is walking through London.

Business took him over London Bridge, and I accompanied him. He was very full of the importance of that book [a bill-book] and magnified it. As we passed over the Thames we paused to look at a steamer unloading great slabs of white and brown marble. A barge drifted under the steamer's stern and a lonely ship's cow in the barge bellowed. Charlie's face changed from the face of the bank-clerk to that of an unknown and—though he would not have believed this—a much shrewder man. He flung out his arm

[96] Rudyard Kipling, ' "The Finest Story in the World" ', *The Bombay Edition of the Works of Rudyard Kipling: Vol. IX, Many Inventions* (London: Macmillan, 1913), p. 77.
[97] Ibid., p. 94.   [98] Ibid., p. 91.
[99] For *The Waste Land* as a poem of metempsychosis see Craig Raine, 'Met Him Pikehoses: *The Waste Land* as a Buddhist Poem', *TLS*, 4 May 1973, pp. 503–5.

across the parapet of the bridge and laughing very loudly, said:—
'When they heard *our* bulls bellow the Skroelings ran away!'[100]

Eliot, like Charlie, was preoccupied with bill books, so that a 'Bill of Lading' even finds its way into his note on 'C.i.f. London: documents at sight'.[101] In *The Waste Land* we move where 'The barges drift / With the turning tide' where, just before, the 'white and gold' marble of the Church of St Magnus Martyr is admired. Crossing London Bridge among the crowd of clerks, Eliot too has a most curious shift from a modern scene to an ancient seaman's cry.

There I saw one I knew, and stopped him, crying: 'Stetson!
'You who were with me in the ships at Mylae!'

Charlie is unaware of the spirits alive in him, so that the narrator of the story is frustrated in his attempt to come at the uncluttered account of past lives.

It remained now only to encourage Charlie to talk, and here there was no difficulty. But I had forgotten those accursed books of poetry. He came to me time after time, as useless as a surcharged phonograph—drunk on Byron, Shelley, or Keats.   ·

Yet Charlie's method also hints at the method which Eliot would use in *The Waste Land*:

The plastic mind of the bank-clerk had been overlaid, coloured, and distorted by that which he had read, and the result as delivered was a confused tangle of other voices most like the mutter and hum through a City telephone in the busiest part of the day.[102]

It is not only that Charlie's lives, like those of Tiresias, grow confused; his expression, speaking out of modern London about remote and contemporary affairs, all in 'a confused tangle of other voices', like a poetic ventriloquist, points in the direction of

> London Bridge is falling down falling down falling down
> *Poi s'ascose nel foco che gli affina*
> *Quando fiam uti chelidon*—O swallow swallow
> *Le Prince d'Aquitaine à la tour abolie*
> These fragments I have shored against my ruins

[100] 'The Finest Story in the World', pp. 94–5.
[101] *CPP*, p. 77, n. on l. 210.
[102] 'The Finest Story in the World', pp. 86–7.

Why then Ile fit you. Hieronymo's mad againe.
Datta. Dayadhvam. Damyata.
Shantih shantih shantih

Charlie Mears does not express himself in Sanskrit, but the narrator
of the text knows who will most appreciate Charlie's story. He tells
it to an Indian friend, and soon ceases to speak the tale in his own
language, substituting instead the Indian's. 'After all, it could never
have been told in English.'[103] The narrator dreams of how Charlie's
tales of his reincarnations will be spread, as 'the finest story in
the world' until the point when, 'Every Orientalist in Europe would
patronize it discursively with Sanskrit and Pali texts.'[104] The narrator
dreams of founding a new religion. *The Waste Land* seems at times
to be attempting to piece together some new religion out of fragments
of the old. It also follows Kipling's story in its juxtaposition of
seamen. Eliot may not have in his poem Kipling's Greek slave on
a galley out of Egypt, but he does give us a slightly earlier sea-
farer who sailed out of the Middle East and whose story might be
thought to be specially appropriate to those clerks who work in
the city —

> Phlebas the Phoenician, a fortnight dead,
> Forgot the cry of gulls, and the deep sea swell
> And the profit and loss.

Overwhelmed by water, Charlie Mears, the London bank-clerk-
cum-Greek-galley-slave met his death by water more than once.
'I had an awful dream about that galley of ours. I dreamed I was
drowned . . . I could hear the water sizzle, and we spun round like
a cockchafer . . . '[105] A little later Kipling presents the description
of a breaking wave as a 'banjo string drawn tight' which so impressed
Eliot. The manner of Charlie's death in this particular incarnation,
coupled with the way in which he is able to look back over his earlier
experiences no doubt contributed a lot to the fate of Eliot's Phlebas,

> A current under sea
> Picked his bones in whispers. As he rose and fell
> He passed the stages of his age and youth
> Entering the whirlpool.

---

[103] Ibid., p. 99.   [104] Ibid., p. 96.   [105] Ibid., p. 89.

Similarly, Charlie's incarnation as a Norse explorer among the Skroelings, ending when a man who seems to be Erik the Red took his crew and 'steered them for three days among floating ice, each floe crowded with strange beasts that "tried to sail with us," said Charlie, "and we beat them back with the handles of the oars" '[106] surely relates to the long sea voyage, an account of which originally formed the bulk of the 'Death by Water' section. In this passage, as in the Kipling story, food runs out and three ghosts haunt the seamen. Norwegian pine is mentioned, and Eliot's voyagers also end among the ice-floes, among polar bears.

Charlie Mears, the bank clerk, however, is more usually a second-rate poet who does not understand the value of his 'dreams', scorning to think of them in terms of actual reincarnations, though these dreams are so vivid that reality and unreality seem mixed. Here was another aspect of the Kipling story which must have appealed to the student of F. H. Bradley, the Eliot who himself wrote on 'Degrees of Reality'.

Small wonder that his dreaming had seemed real to Charlie. The Fates that are so careful to shut the doors of each successive life behind us had, in this case, been neglectful, and Charlie was looking, though that he did not know, where never man had been permitted to look with full knowledge since Time began.[107]

This vision, what Eliot called Kipling's knowledge 'of the things which are underneath, and of the things which are beyond the frontier', was what made some of Kipling's short stories so important to the writer of *The Waste Land*.[108] It was like Frazer's looking into the 'abysm of time', but it was a vision only imperfectly appreciated by Charlie Mears, as the narrator of ' "The Finest Story in the World" ' emphasizes: 'Above all, he was absolutely ignorant of the knowledge sold to me for five pounds; and he would retain that ignorance, for bank-clerks do not understand metempsychosis, and a sound commercial education does not include Greek.'[109] To find its most lasting realization, Charlie's vision would have to wait for a bank clerk who did understand metempsychosis, and who had not only a commercial education gained in an underground room at Lloyds Bank, but also a knowledge of Greek. *The Waste Land*'s tone would be grimmer, 'throbbing between two lives'. But when such a bank clerk followed Charlie, Rudyard Kipling entered *The Waste Land*.

---

[106] 'The Finest Story in the World', p. 107.   [107] Ibid., p. 86.
[108] *OPP*, p. 239.   [109] 'The Finest Story in the World', p. 86.

Kipling provided links between life in the modern city and ideas of incarnation, but Eliot further bound up these two elements with primitive ritual. Again, Eliot found artistic incitements for doing this. His 1923 review of *Ulysses* ends by stating that a combination of psychology, ethnology, and Frazer's *Golden Bough* has made possible a new method of artistic construction. 'Instead of narrative method, we may now use the mythical method.' But it would be wrong to see *The Waste Land* as 'controlling . . . ordering . . . giving a shape and a significance to the immense panorama of futility and anarchy which is contemporary history' in the way that Eliot believed *Ulysses* did.[110] For *The Waste Land*, while it does unite the themes which Eliot sees as paramount in 'contemporary history' and history as a whole, gives to those themes no firm significance which raises them above futility.

The tribes of Todas and Veddahs stuck in Eliot's mind for at least fifteen years because both had little sense of the meaning of their religious ceremonies.[111] At Harvard he read how apart from gods useful for economic purposes, 'All the myths about their gods are otherwise hazy, and their names persist in a meaningless way in the Todas' prayers.' Such a state of affairs is related to that of 'The Kafirs of South Africa' who have, we are told, 'no definite social structure' and whose 'notion of Umkulunkulu, one of their chief divinities, [is] . . . extremely hazy, and there is little agreement as to who he really is.'[112] Eliot mentioned Umkulunkulu in his 1913 paper.[113]

Ideas of the relationship between developed drama and primitive rites and of the loss of meaningful mythologies continued to haunt Eliot. In spring 1921, when 'the English myth is pitiably diminished' he discussed 'the chief myth which the Englishman has built about himself', that of the fat country squire. The language used to describe this is reminiscent of Harrison, as well as clearly relating to the technique of Charlie Mears, Tiresias, and the *dromenon*-like patterning with which Eliot had been experimenting in the quatrain poems.

The myth that a man makes has transformations according as he sees himself as hero or villain, as young or old, but it is essentially the same myth; Tom

---

[110] 'Ulysses, Order, and Myth', *Dial*, Nov. 1923, p. 483.
[111] See, e.g., 'Introduction' to *Savonarola* (B4), p. x [for Todas, sc. Veddahs] and *SE*, p. 44.
[112] King, pp. 236 and 97. [113] IPR, p. 3.

Jones is not the same person, but he is the same myth as Squire Western; Midshipman Easy is part of the same myth; Falstaff is elevated above the myth to dwell on Olympus, more than a national character.[114]

As in 'Exequy', Eliot ends with a ridiculous man-god. His piece, appearing in *Tyro* beside Lewis's totem-pole-like modern figures and John Adams's 'Cafe Cannibale', was part of a contemporary interest in primitive roots. *Ulysses* was for Eliot in 1921 'the greatest work of the age'.[115] Probably the most important effect of his reading of Joyce was to make him all the more aware of the possibilities of his anthropological reading, especially when applied to modern city life in the context of inanity or death: city life was filled with fatal torpor and Eliot described London as shrivelling, like an aged little bookkeeper.[116] In November 1921 Eliot expressed a dread of London, longing for sea or mountains.[117] In his discussion of the decay of English myth, he holds out only the slender hope of the music hall and laments that in general modern dramatists and probably modern audiences are 'terrified of the myth'.[118] Eliot at this period was searching amongst myths and rituals. Weston's 1920 *From Ritual to Romance* gave him the tarot pack and backed up Frazer's emphasis on links between sexuality and religion. She furnished Eliot with the title and central concept of his poem, but was otherwise more important for summarizing and reminding him of other people's ideas, such as those of E. K. Chambers speaking of literary forms encompassing 'fragments of forgotten cults'.[119] Through quotation and allusion Weston reminded Eliot of his reading in Cornford, Frazer, Harrison, Spencer and Gillen, and Gilbert Murray, and discussed many of the questions which Eliot had explored in their pages. His mind filled with primitive lore and with a sense of awkwardness at the numerous exhausting social roles he had to play in addition to that of the London banker, Eliot wrote to Mary Hutchinson in 1920 worrying about his inherited characteristics and suggesting that he might be a savage himself.[120] At

---

[114] As n. 82 above.
[115] *The Question of Things Happening; The Letters of Virginia Woolf, Volume II: 1912–1922*, ed. Nicolson and Trautman (London: The Hogarth Press, 1976), p. 485 [letter, to Roger Fry, 17 Oct. 1921].
[116] 'London Letter', *Dial*, May 1922, p. 510.
[117] Letter to Schiff [*c*.6 Nov. 1921] (British Library).
[118] As n. 82 above.   [119] Chambers quoted in Weston, p. 66.
[120] Letter to Mary Hutchinson, 20 Sept. 1920 (Austin, Texas).

home, he read in Weston's book that

In the *Rig-Veda* . . . Indra, while still retaining traces of his 'weather' origin, is no longer, to borrow Miss Harrison's descriptive phrase, 'an automatic explosive thunder-storm,' he wields the thunderbolt certainly, but he appears in heroic form to receive the offerings made to him, and to celebrate his victory in a solemn ritual dance.[121]

Given the various elements present in Eliot's mind, it is hardly surprising that he found in the thundering drums of Stravinsky's ballet, *Le Sacre du printemps*, the equivalent of the myth he sought. Attending the ballet, Eliot sought to use the point of an umbrella in order to stop his neighbours laughing.[122] The city man's rolled umbrella could easily become a spear. For Eliot Stravinsky's piece itself united city and savage, seeming to metamorphose 'the rhythm of the steppes into the scream of the motor horn, the rattle of machinery, the grind of wheels, the beating of iron and steel, the roar of the underground railway, and the other barbaric cries of modern life; and to transform these despairing noises into music'.[123] This is hardly an average reaction to the piece, but the despairing Eliot was using Stravinsky's music to underpin his own endeavours in *The Waste Land*. A member of Diaghilev's company had become Grishkin, uniting jungle and 'maisonette', in 'Whispers of Immortality'.[124] Eliot's reading of *Heart of Darkness* whose Buddha-like clerk, Marlow, saw London as 'one of the dark places of the earth' further blended savage and 'sepulchral city'.[125] Eliot returned to the Russian Ballet in 1919, and in 1921 saw their dancing in terms of vegetation rites and primitive ceremonies, but that dancing contained only marginal interest because not linked to the modernity of Stravinsky's music.[126] It was, however, of interest to those who had read *The Golden Bough* and related works. 'In art there should be interpenetration and metamorphosis. Even The Golden Bough can be read in two ways: as a collection of entertaining myths, or as a revelation of that vanished mind of which our mind is a

[121] Weston, p. 36.   [122] 'A Commentary', *Criterion*, Oct. 1924, p. 5.
[123] 'London Letter', *Dial*, Oct. 1921, p. 452.
[124] Serafima Astafieva—see *Ezra Pound and Music: The Complete Criticism*, ed. R. Murray Schafer (London: Faber & Faber, 1978), p. 482.
[125] Joseph Conrad, *Heart of Darkness* (1902; rpt. London: Dent, 1974), pp. 50, 48, 152.
[126] Vivienne Eliot, 1919 diary (Bodleian); as n. 123 above.

continuation.'[127] That last phrase, presenting the savage mind as 'vanished', yet suggesting that it continues in our own, is in line with Eliot's refusal when discussing Lévy-Bruhl, to pin down just how the savage mind is related to the modern. In his poem it seems related to modern city life by both 'interpenetration and metamorphosis'. Weston emphasized to Eliot that not only dramas, but poetry also derived from primitive rites. By adopting the Waste Land theme, Eliot made himself the performer of the latest enactment of a repeated rite, just as in his poem clerk and typist (like the earlier Burbank and Volupine) enact their own sexual ritual. Though doubtless partly motivated by a desire to draw the reader away from his own personal life, Eliot was serious in pointing out that, 'Anyone who is acquainted with these works [of Weston and Frazer] will immediately recognize in the poem certain references to vegetation ceremonies',[128] which, as in the quatrain poems, were associated with religion and human sexuality. But in the same way that the sexual act is losing its meaning, in pub scene, or baroque interior, or city flat, with the result that we are brought to question whatever meaning it once may seem to have possessed, so in *The Waste Land* the rituals of religion appear also to be failing.

Eliot had recounted in 1916 how in the excitement of the Australian aboriginal corroboree 'with every stimulant of noise, torchlight, strange masks, and drink, the savage seems to himself to have become a new being'.[129] Eliot was relying on Durkheim's account of *intichiuma* ceremonies with 'all sorts of processions, dances and songs . . . by torchlight . . . a . . . savage scene'.[130] Durkheim went on to link aboriginal with Christian practices. Eliot's earlier poetry had endowed scenes of Christian martyrdom with a savage violence and primitive overtones. Now he linked corroboree, *intichiuma* of spring, and Crucifixion, as if to find beneath the 'stony places' of the Bible the 'stony places' of the Australian desert photographed by Spencer and Gillen and presented in *The Elementary Forms of the Religious Life*.

> After the torchlight red on sweaty faces
> After the frosty silence in the gardens
> After the agony in stony places

---

[127] As n. 123 above, p. 453.   [128] CPP, p. 76.
[129] 'Durkheim'.   [130] Durkheim (1971), p. 218.

The shouting and the crying
Prison and palace and reverberation
Of thunder of spring over distant mountains
He who was living is now dead
We who were living are now dying
With a little patience

'Dying' because the sustaining myths seem to be fading away. There
is deliberate incongruity in the cry to Stetson. In *The Waste Land*
the idea of a fertility cult in City or suburb is made to seem not
irrelevant, but horribly absurd as in the scenes overheard in the pub
or observed by Tiresias. Primitive religion interpenetrates other
material throughout a poem which opens with roots, plant life, the
stuff of heredity and apparently the basis of the most elementary
religion. In a work so concerned with the transmission of cultural
values this is appropriate. Roots and branches speak of plant-life and
of the gods which anthropology stressed were totemically associated
with it. They introduce the 'Son of man', and the relationships
between Christianity and other religions whose heaps of broken
images strew the anthropologists' pages and the sites of the world's
past cultures. Other themes, such as the importance of the shadow
to primitive man, flicker through these lines. That memorable
'handful of dust' belongs to the religion of 'ashes to ashes, dust to
dust' in this burial of the dead, but also to more primitive rites. As
April, 'the cruellest month', brings renewed life, which is a renewal
of pain and memories of destruction, so antithetically, this dust looks
not only towards the grave, but also towards birth as the handful
of fecundating dust sprinkled over those red rocks of Durkheim's
desert tribes.

. . . the grains of dust which the Australian detaches from the sacred rock
are so many sacred principles which he scatters into space, so that they may
go to animate the totemic species and assure its renewal.

The Waste Land itself functions as a primitive ritual. Cries, Eliot
knew, were vital to the most basic corroboree when 'on every side one
sees nothing but violent gestures, cries, veritable howls, and deafening
noises of every sort'. In religious ceremony, 'the same manifestations
are to be observed in each case: cries, songs, music, violent movements,
dances, the search for exciteants which raise the vital level'.[131] Such

[131] Ibid., pp. 342, 216, and 383.

musical cries 'raise the vital level' in Eliot's poem, bursting strangely into the texture of the English:

> The peal of bells
> White towers
> > Weialala leia
> > Wallala leialala

But other cries are also present. Spencer and Gillen photographed and detailed aborigines dressed as animals, emitting animal cries to promote fertility. We may have a glimpse of sympathetic magic in 'when we came back, late, from the hyacinth garden, / Your arms full, and your hair wet . . . '. Jessie Weston reinforced Frazer in stressing that 'Throwing into, or drenching with, water is a well-known part of "Fertility" ritual; it is a case of sympathetic magic, acting as a rain charm.'[132]

But *The Waste Land*'s sounds, rather than sights, come closest to enacting savage ritual, as we might expect of Eliot's 'auditory imagination'. Writing of human imitation of totemic creatures, Durkheim stressed that 'In a large number of tribes, the Intichiuma of rain consists essentially in imitative rites.' Among the Arunta 'in the Intichiuma of water, the men of the totem utter the characteristic cry of the plover, a cry which is naturally associated in the mind with the rainy season.'[133] Eliot gives us not plovers' sympathetic magic but that of the *Turdus aonalaschkae pallasii* whose ' "water-dripping song" is justly celebrated'.[134]

> Drip drop drip drop drop drop drop
> But there is no water

The ritual now is powerless. The magic fails. *The Waste Land*'s bestiary is a full one; other creatures such as bats and crickets were associated with rainmaking. As Lévy-Bruhl pointed out, primitive men saw the chirping of crickets and crying of birds in spring as appeals for rain.[135] Such cries traditionally used to promote fertility occur ironically in *The Waste Land*, punctuating its infertile or sexually perverted world where previous rituals and beliefs seem to be lapsing into futility.

---

[132] Weston, p. 51 n. 31.   [133] Durkheim (1971), pp. 352–3.
[134] *CPP*, p. 79.   [135] Lévy-Bruhl, pp. 251–2.

Twit twit twit
Jug jug jug jug jug jug

. . .
Co co rico co co rico

The first cry is associated with 'The change of Philomel, by the
barbarous king / So rudely forced'. Eliot's interest in human animals,
seen in the quatrain poems, intensifies in *The Waste Land*. It is
hardly surprising that a passage of Ovid is described in his Notes
as 'of great anthropological interest'.[136] The poem's most crucial
cry is that of the cock. It comes from Rostand's *Chantecler*, a play
which Eliot recalled in 1919, probably having seen it in Paris where
it had been a hit in the winter of 1910.[137] This play which was in
Boston in 1911 is set in a farmyard. The actors dress in animal or
bird costumes.[138] A nightingale is one of the chief characters. The
play must have struck Eliot as fitting in perfectly with *intichiuma*
ceremonies and with the general idea of drama evolving from
primitive ritual. Vain Chantecler refuses to recognize that it is not
his own crying of 'Cocorico' which causes the sun to rise. His
description of that cry is, to anyone interested in anthropology, heavy
with overtones of fertility ritual: 'La Terre parle en moi comme dans
une conque.'[139] *Chantecler* must have fuelled Eliot's convictions
about the origins of drama in primitive ceremony, as well as dealing
with problems of religious decay and with mystical *versus* rational
belief. Watching actors in animal costume making their sophisticated
jokes was like proof of anthropologists' statements, as typified by
Frazer in that same conclusion to *The Dying God*, where he stated
that serious rites had become idle amusement, mummeries, and
childish games.[140]

Eliot, who occasionally attended Ouspensky's seances in London
in 1920 and who used the Tarot and Madam Sosostris in the London
of *The Waste Land*, was familiar with the decay of magic; but the
man who attacked *Back to Methuselah* in 1921 hardly shared Frazer's

[136] *CPP*, p. 78.   [137] See Timothy Materer, 'Chantecler in "The Waste
Land" ', *Notes and Queries*, Oct. 1977, p. 451.
[138] 'Chantecler at the Hollis Street [Boston]' was reviewed in *Harvard Advocate*,
92 (Nov. 1911), p. 72.
[139] Edmond Rostand, *Chantecler* (Paris: Librairie Charpentier et Fasquelle,
1910), p. 108.
[140] Frazer, *The Dying God*, ed. cit., p. 269.

Comtian, progressive tone. Shaw's book, however, did encourage
Eliot to think of the modern in terms of the ancient, of 'creative
evolution', of 'Darwin', 'biology', and the 'ultimate questions' raised
by 'the attempt to expose a panorama of human history "as far as
thought can reach"'.[141] 'Social evolution' and its application to
modern English society were already in Eliot's mind. He attacked
contemporary literature's taste for which all things were either too hot
or too cold, and complained that it contained no culture. He lamented
the extinction of the music hall, home of the vestiges of the English
myth.[142] *The Waste Land*, in presenting extreme emotional states,
would encompass the extremes of city and savage. Durkheim and
others had emphasized to Eliot connections between primitive rituals
like the *intichiuma* and more developed religions such as Christianity.
For Durkheim and his associates such ceremonies did not represent,
as they did for Frazer, magic as some separate predecessor of religion,
but were themselves elementary forms of the religious life. *The Waste
Land* hardly suggests that the inhabitants of the twentieth-century
city are conscious actors in fertility rites, but since these, long
forgotten, underlie our behaviour, since the 'sexual instinct' plays
a role in 'the religion and mythology of primitive peoples (indeed
in all religion)', and since Christianity and primitive ritual are linked,
the poem expresses despair at the change and decay not only of city
churches which Eliot visited at this time, but of all belief.[143]

Yet under all the various forms, the sexual act and natural fertility
cycles persist. In cities, where the seasons' impact is dulled, the rituals
of fertility seem to lose their meaning, but they continue, processing
like scenes in a play before Tiresias, epitome of the sexual process,
male and female trapped together, who has 'foresuffered all'. These
modern performances interpret all earlier performances and interpretations, casting doubt on their total significance, just as, on a
smaller scale, Eliot's reworking of the Enobarbus speech casts doubt
on the validity of Shakespeare's interpretation of Cleopatra.

Whether Carthage or London, the poem's cities are seen as
horrible, life-denying.

---

[141] Behr (n. 68 above), p. 20; 'London Letter', *Dial*, Oct. 1921, pp. 453–5.
[142] 'London Letter', Apr. 1921, p. 451 and 453; see also 'London Letter', *Dial*,
June 1921, pp. 687–8.
[143] Review of Wundt, *Monist*, Jan. 1918, p. 160; on the destruction of city
churches see 'London Letter', *Dial*, June 1921, pp. 690–1 and *CPP*, p. 78 and
Facsimile, p. 128.

> Jerusalem Athens Alexandria
> Vienna London
> Unreal

Yet the primitive pattern of sexual 'burning' underlying their inhabitants' actions and beneath religious belief is scarcely more attractive.

If the poem goes back to the origins of religion, it also goes back to the origins of society and language. The old order has been destroyed. *The Waste Land* begins a new poetry, or seeks to, but also marks the reborn post-war society. The lines which read in draft

> Who are those hooded hordes swarming
> Over Polish plains, stumbling in cracked earth

became in the final version universalized through the substitution of 'endless' for 'Polish'. The constant 'hordes' is significant. Eliot knew the technical sense of this word: 'The *horde* . . . is the protoplasm of the social domain. . . .'[144] Wundt wrote that about '*horde*, meaning . . . an unorganized, in contrast to an organized, tribe of people . . . The Tartars called a division of warriors a *horda*.'[145] The horde, 'Characteristic of primitive times', was the basic social unit. Eliot, who was putting 'Tartar horsemen' in his poetry in 1924, seems to have been impressed by this passage.[146] *The Waste Land* returns to all sources, including the sources of human society.

What it brings back seems little, in terms of hope at least. Religious hope flickers, as at Magnus Martyr, but the city crowds out the city churches, and going to the apparent sources of the religious life which seemed to hold promise in Lower Thames Street, Eliot found himself unable to sustain hope. 'Eastern and western asceticism' merge.[147] Christian Burial of the Dead combines with Buddha's Fire Sermon, but both are shot through with the sort of primitive fertility cycle hinted at in 'Death by Water'. The poem's exploration reveals all religions combined. We are deliberately led on towards what Eliot knew was one of the phenomena found most universally in religions ranging from the most primitive to the most developed: thunder.

The veneration of thunder was mentioned continually in Eliot's anthropological reading: thunder as the terrifying voice of a god,

---

[144] Durkheim (1982), p. 113.
[145] Wundt, op. cit., p. 52.
[146] *CPP*, p. 134.
[147] *CPP*, p. 79.

commanding respect and faith.[148] Such a voice was necessary also, as bringing the rain so essential to life's continuation. Among the Australians, Harrison wrote, sounds of thunder were simulated.

The roaring, boys and women are told, represents the muttering of thunder, and the thunder—this is the important point—is the voice of Dhuramoolan. 'Thunder,' said Umbara headman of the Ywin tribe, 'is the voice of Him (and he pointed upwards to the sky) calling on the rain to fall and everything to grow up new.'

Thunder was associated with the Revelation story of death and rebirth (see, for example, Rev. 8: 5), with Australian rainmaking, with the Grail story lightning, and with the origin of drama. Concerning thunder rites, Harrison even went so far as to quote Durkheim's *'Le sacré, c'est le père du dieu.'*[149] From sacred thunder the voice of a god might be born, a desert might be transformed, and man, savage man and even modern City man, might live again.

Ideas of the juxtaposition of the arid and the flowering desert were familiar to Eliot from the time when he had read Mayne Reid. They grew with his reading of Frazer who described the 'Burnt Land of Lydia' contrasting with the surrounding verdure and marvelled (in the conclusion of *The Dying God*) at 'what may be called the Australian spring' where 'the sandy and stony wilderness, over which the silence and desolation of death appear to brood, is suddenly, after a few days of torrential rain, transformed into a landscape smiling with verdure'.[150]

Thunder, bringing rain, God's saving word, made possible 'a new start'. To find this, Eliot goes from the city to the savage beginnings of religion, from literature and drama to their ritual origins, and from his present as a London clerk to his earlier life. There he found fragments uniting the personal and anthropological, whether in the 'memory and desire' of the Thomsonian buried corpse about which he had read at Harvard, in the Frazerian Mayne Reid deserts of his childhood, in Kipling's metempsychosis, Rostand, or Jacobean dramatists, or a passage recommended to him by his Harvard Sanskrit teacher, Charles Lanman, who had laid special emphasis on the advice which the Hindu 'Lord of Creatures' gives to men in

---

[148] See, e.g., F. M. Cornford, *The Origin of Attic Comedy* (London: Edward Arnold, 1914), p. 23; Frazer, *Adonis Attis Osiris*, i. 134; Harrison, p. 61 and ch. III *passim*.

[149] Harrison, p. 63.   [150] Frazer, *The Dying God*, p. 270.

thunder. In a letter preserved in Eliot's copy of the *Upanishads*, Lanman mentions this as an important passage.[151] But in Eliot's poem the response to these commands is uncertain and generally one of failure. The 'awful daring of a moment's surrender' is an experience intense, but imprudent. All there is in this thunder is not relief but a revival of 'aethereal rumours' which 'Only at nightfall . . . / Revive for a moment a broken Coriolanus.' Hope is fleeting ('for a moment') and lost ('broken'). 'Your heart *would have* responded', but did not (my italics).

The thunderword appears as the solution, the rain, a new beginning. It looks back to Lanman as well as to Eliot's other Sanskrit teacher, the ex-anthropologist, Woods, who had written that to the Hindu, 'Logically and temporally, the word seems to precede its idea and its meaning: a man thinks, the Hindu would say, because he is talking. The benefactor of the race would be, not a Prometheus who brings a few sparks from above, but he who releases among men the most finished of all forces, an irresistible word.'[152] It is not difficult to relate such an idea both to Eliot's general preoccupation with articulation in his poetry, and in particular to *The Waste Land* where Eliot includes Lanman's explanatory gloss, 'da — da — da = *damyata datta dayadhvam*',[153] in his text. After exhausting the gamut of expression from Cockney pub-talk to Dante, and running across the broad acres of comparative religion from *intichiuma* to St Magnus Martyr, Eliot seems to be generating the commands of a new religion reborn from the old, by returning in his rainmaking to the origin of religious rites. He also appears to be physically generating a new language, having gone back to the simplest animal and natural cries. We hear the words being formed from nature:

> DA
> Datta
> . . .
> DA
> Dayadhvam
> . . .
> DA
> Damyata

[151] See C. R. Lanman, Course Records 1892–1926, p. 73 (Pusey) and Robert Bluck, 'T. S. Eliot and "What the Thunder Said"', in *Notes and Queries*, Oct. 1977, pp. 450–1.
[152] Woods (ch. III n. 55 above), p. 91.   [153] As n. 151 above.

This would seem very close to the 'bow-wow' theory of linguistic evolution, one of the theories much discussed in Eliot's youth, which saw language as deriving from the imitation of natural sounds. Such a theory had been largely abandoned. Certainly to derive the polysyllabic *Dayadhvam* from the monosyllable DA is a precarious process. Particularly associated with the discussion of theories of the linguistic origins of divinities was Max Müller, whose views Eliot saw as outdated.[154] Praising Durkheim, Eliot had written about and summarized Max Müller's attitude (largely based on Sanskrit philology), which looked at primitive religion chiefly through its myths, finding in primitive deities personifications of larger forces of nature and seeing in the primitive mind a sentiment of wonder at such forces, a basic intuition of the 'infinite'. But Durkheim, Eliot noted, did not attribute the origin of religion to wonder or to speculation, but saw in mythology only the savage's attempt to rationalize and justify his religious practices, 'in regard to the true origin of which he is as much in the dark as the scientific investigator'.[155]

The Waste Land is a poem which leaves its readers in darkness. The rituals of the past are perverted and decayed. Savage and city alike reduce to childishness in the final babel-like collapse, 'London Bridge is falling down falling down falling down.' The thunder ritual, which might offer salvation, is deeply suspect, potentially inane. It is yet another part of the cycle of fertility which is seen in *The Waste Land* as a torturing cage. The poem is not structureless. Themes discussed above clearly unite the diverse materials. If the poem were a formal mess, reordering the five sections would make no difference; in fact, such a reordering would destroy the cyclic form. Even if the awaited rain fell at the poem's end, it would only lead back to that beginning, 'breeding / Lilacs out of the dead land,' with all its attendant suffering. Eliot in his July 1921 'London Letter' saw the result of a hot rainless spring simply as a 'poor' 'crop of murders and divorces' in a 'vacant term'.[156] Whitman's sentiments

> When lilacs last in the dooryard bloom'd,
> . . .
> I mourn'd, and yet shall mourn with ever-returning spring.

---

[154] See Brian Stross, *The Origin and Evolution of Language* (Dubuque, Iowa: Wm. C. Brown, 1976), pp. 20–24.
[155] 'Durkheim'.   [156] 'London Letter', *Dial*, Aug. 1921, p. 213.

have been taken over with a vengeance and put into the anthropological perspective which in the poem Eliot has applied to so much past literature. Life is painful throughout, with only the briefest moments of relief—the hyacinth garden, the fishmen at Magnus Martyr, boating—all of which only lead back into the cycle of further (often sexual) pain. The 'Shantih' at the poem's end may be simply a way of stopping, a 'formal ending';[157] or it may be comparable to the exhausted collapse after the destruction at the end of 'Gerontion' (Eliot considered 'Gerontion' as preface to *The Waste Land*).[158] Savage and city, bound together in a world fundamentally meaningless and hurtful, furnish expression for Eliot's private torment. If there was to be an escape, it was to lie not in the 'damp gust bringing rain' but in 'Shantih' as the Buddhist's passing beyond the cycles of creation.

Eliot was unable to renounce the world he knew. He remained with Vivienne, the bank, and years of exhaustion in which he came close to death. He decided eventually that to embrace Buddhism would be to plunge into a world too culturally alien. In the time following *The Waste Land* he saw the world despairingly, in terms of inane, savage horror. Run-down primitive rituals stalk 'The Hollow Men'; in *Sweeney Agonistes* they overpower jazz-age London. That play has as an epigraph a Christian equivalent of the escape through 'Shantih' from the cycles of creation: '*Hence the soul cannot be possessed of the divine union, until it has divested itself of the love of created beings.*'[159] But *Sweeney Agonistes* follows *The Waste Land* in presenting a world where fertility, renewing painful, inane life, is inescapable. To see more deeply is only to see what Eliot found in the epigraph from *Heart of Darkness* which he thought might have been 'somewhat elucidative' for *The Waste Land*.[160] Conrad too found the primitive and metropolitan uniting to hint at something underlying and awful, ' "The horror! the horror!" ' Tortured between the city and the savage, Eliot's was a nightmare world.

---

[157] *CPP*, p. 80.
[158] Pound (1950), p. 171, Eliot to Pound, London [?Jan. 1922].
[159] *CPP*, p. 115.   [160] As n. 158 above.

V

THAT'S ALL, THAT'S ALL, THAT'S ALL, THAT'S ALL

IN any marshalling of detailed evidence the dangers of forensic tedium are present. They must always be risked. Clues, however remote, may be crucial. The argument for preserving Shakespeare's laundry bills is that, eventually, they may find their Sherlock Holmes.

The reader should bear in mind Eliot's exasperation at the graduate student who 'identifies Corbière's "Rhapsode Foraine" as an exploration "of folk-religion" '.[1] Nevertheless, in the early twenties Eliot's mind was clearly full of not only 'Frazer . . . Durkheim . . . Lévy-Bruhl', but also of folklorists and anthropologists such as Sir E. B. Tylor, Robertson Smith, Miss Harrison, Miss Weston, A. B. Cook, F. M. Cornford, le Dr. Rendel Harris, Hartland, Elliot Smith, Gilbert Murray, et d'autres encore,—pour ne rien dire d'observateurs tels que Codrington, Spencer et Gillan [sic] Hewett, [sic] qui ont étudié les races primitives aux quatre coins de l'Empire.'[2] Eliot's spelling of 'Gillen' and his misremembering of Howitt's surname show that he was recalling material from his past. In 'The Hollow Men' and *Sweeney Agonistes* we see that such material had not lain gathering dust. It had germinated.

'The Perfect Critic' (1920) discusses Arnold in the context of English criticism as dead or dying.[3] In a letter of that autumn Eliot confessed to using Arnold as a scarecrow—apparently lifeless, but having under his arm a real gun to be fired against the old guard.[4] Later he saw Arnold as initiating 'the degradation of philosophy and religion'. Associations of the scarecrow with powerlessness and religious degeneration resurface in 'The Hollow Men', where the scarecrow is again camouflage, but this time hopelessly unarmed:

---

[1] 'Contemporanea', *Egoist*, June/July 1918, p. 84.
[2] 'Lettre d'Angleterre', *Nouvelle Revue Française*, 1 Nov. 1923, p. 622.
[3] *SW*, p. 1.
[4] Letter to Sidney Schiff, marked simply 'Friday', but dated as 1920 in British Library. For Arnold and 'degradation', see *SE*, p. 437.

> Let me also wear
> Such deliberate disguises
> Rat's coat, crowskin, crossed staves
> In a field
> Behaving as the wind behaves

The crossed staves of Christian iconography become a place for hanging vermin. Dressing in animal skins for ritual purposes and making divine effigies from straw are favourite Frazerian topics, as is 'The Propitiation of Vermin by Farmers'.[5] The epigraph, '*A penny for the Old Guy*', stresses that Eliot's poem relates to ceremonial effigies. Here, as in *The Waste Land*, *Heart of Darkness* is important. '*Mistah Kurtz—he dead*' emphasizes a connection between savage ritual and Eliot's crossed staves. To obtain power by embracing darkness, Kurtz deified himself in line with primitive belief; ironically, Eliot's speaker dresses in relics of forgotten ritual out of a sense of total impotence, wishing to avoid a horrid dusk: 'Not that final meeting / In the twilight kingdom.' It is landscape like Dowson's, 'Hollow Lands' where, in 'the twilight of the year', 'dead people with pale hands / Beckon' by a 'weary river', 'where pale stars shine' where, at passion's enactment, 'There fell thy shadow'.[6] But the poem's world, lit by the half-light of inaction, was a transmutation of Eliot's own suffering. Vivienne, who had felt *The Waste Land* to be a part of herself, saw 'The Hollow Men' as a fitting follow-on, and related it to her own nightmares.[7] But, as before, the personal world and that of late nineteenth-century poetry are sieved through anthropological ideas, emotion distributed among unsettled voices, narrative, optative, and choral. The poem is typographically complex. The first section's 'Kingdom' may not be section three's 'kingdom'. Certainly 'death's *other* kingdom' (my italics) suggests that the speakers' inane life is only another form of death. So, for Sweeney Agonistes 'Death or life or life or death / Death is life and life is death.' Cornford had quoted the Euripidean question, '*who knows if to be living be not death?*'[8] Through anthropology Eliot

[5] J. G. Frazer, *Spirits of the Corn and of the Wild*, ii, ch. XV.

[6] Dowson, *Poetical Works*, pp. 127 ('A Last Word'), 68 ('Autumnal'), 109 ('In a Breton Cemetery'), 54 ('Vanitas'), 65 ('Beata Solitudo'), 52 ('Non Sum Qualis Eram Bonae Sub Regno Cynarae').

[7] Vivienne Eliot, letters to Violet Schiff, 'Monday' [1922] and London, 31 Mar. [1924] (British Library).

[8] Cornford, p. 82.

resuscitates the idea of the living dead, so common in late romantic poetry, where Death-in-Life is the eternal king. Eliot's Hollow Men share their hopelessness with the inhabitants of the City of Dreadful Night.

Though partaking of classical underworlds, Eliot's kingdoms draw heavily on other kingdoms of the dead. Lévy-Bruhl, citing Ellis's description of West African wandering spirits, used the phrase 'dead land' (*le pays des morts*).[9] His 'desert soul' points to Eliot's 'cactus land'. For Lévy-Bruhl's savage the 'invisible . . . is most real' and the dead 'occasionally . . . heard in the wind' when, ghostly, they blow straws in the air; 'the murmuring of the wind betrayed the presence of the dead'.[10] Eliot's 'voices . . . In the wind's singing' signal a dreaded presence:

> Eyes I dare not meet in dreams
> In death's dream kingdom
> These do not appear:
> There, the eyes are
> Sunlight on a broken column
> There, is a tree swinging

Oesterley's *The Sacred Dance*, read by Eliot in 1923, discussed a savage who was awestruck when 'a tree, swayed by the wind, moved'.[11] The 'broken column' may relate to the traditional grave-yard image of premature death, as well as to the 'broken stone' of section three where 'stone images / Are raised'. In W. J. Perry's *The Origin of Magic and Religion* which he reviewed in July 1924, four months before these lines appeared, Eliot read of 'old stone images' of the Melanesians, of *mana*, of the handing down of rituals, and of W. H. R. Rivers's work on 'an extensive literature in which attempts are made to bring the symbolism of myth and ritual into relation with modern views concerning its rôle in the dream and disease'.[12] For Eliot in 1922, Rivers had been 'the great

---

[9] Pointed out in Marc Manganaro, 'Lost Connections: T. S. Eliot and Lucien Lévy-Bruhl', unpublished paper delivered before a section of the MLA at Emory University, Atlanta, Georgia, during Oct. 1983. See Lévy-Bruhl, p. 83.

[10] Ibid., pp. 88 and 302.

[11] W. O. E. Oesterley, *The Sacred Dance: A Study in Comparative Folklore* (Cambridge: Cambridge University Press, 1923), pp. 14–15. Reviewed in 'The Beating of a Drum' (C146).

[12] W. J. Perry, *The Origin of Magic and Religion* (London: Methuen [1923], pp. 168, 167, 183.

psychologist'.[13] But, in Perry, he also read how the worship of stones fed off an earlier civilization and was dying out. Such reading must have recalled Eliot's Harvard reading in anthropology and psychology of religion when he had noted, amongst other things Jevons's stress that 'the worship of stones is a degradation of a higher form of worship'. Jevons also stated that 'ceremonies may continue to be performed as a matter of custom and tradition long after their original purpose and object have been forgotten'.[14] Eliot's stone images which 'receive / The supplication of a dead man's hand' epitomize the dying embers of rites reduced to meaninglessness.

Such ritual brings no hope, and it diverts to barrenness emotions which might otherwise have been fruitful.

> Is it like this
> In death's other kingdom
> Waking alone
> At the hour when we are
> Trembling with tenderness
> Lips that would kiss
> Form prayers to broken stone.

Again sex and religion combine through anthropological images invested with pain. 'Art and religion', Harrison had stated, 'alike spring from unsatisfied desire.'[15]

The poem concerns degradation of language and ritual, failings of word and Word. Eliot's witty 1918 truncation of an Arnoldian phrase prefigures the Hollow Men's predicament; Clive Bell, 'lingering between two worlds, one dead', has the vice of inane mediocrity, incapable of 'icy inviolability, or violent fury'. His being 'the Matthew Arnold of his time'[16] again connects, through the scarecrow image, with the Hollow Men who, like so many of Eliot's literary adversaries, are afflicted with a stuffed language, emptily 'poetic', the degradation of great poetry—

> Leaning together
> Headpiece filled with straw. Alas!

After the starkly factual 'Headpiece filled with straw' comes that most absurdly poetic word of all, indicating the speaker's weakness. Such

---

[13] 'London Letter', *Dial*, Dec. 1922, p. 662.
[14] Jevons (ch. III n. 52 above), pp. 141 and 254.
[15] Harrison, p. 44.   [16] 'Shorter Notices', *Egoist*, June/July 1918, p. 87.

childish poeticizing is reinforced, in directing us to the level of the infant, by the '*penny for the Old Guy*' epigraph, by the dressing up as a scarecrow, and by the nursery rhyme, 'Twinkle twinkle little star', which inescapably underlies the line 'Under the twinkle of a fading star'. This use of nursery rhyme looks back to *The Waste Land* with its 'London Bridge is falling down falling down falling down' and anticipates another explicit nursery rhyme which, in slightly distorted form, opens section V of 'The Hollow Men'. Other forms of words instilled into the young are also present. The Lord's Prayer is quoted in the repeated fragment '*For Thine is the Kingdom*' but the earlier uses of the word kingdom, particularly when it has a capital 'K', must already have prompted questions about allusion to this fundamental Christian rite, and about how it relates to those 'prayers to broken stone'. It is hard to avoid thinking that the hope of 'Thy Kingdom come' finds its opposite in

> Let me be no nearer
> In death's dream kingdom

which becomes an anti-Lord's Prayer, addressed to an unnamed hearer who may be the god of a universe of barrenness and death. 'The Hollow Men', not 'The Hippopotamus', is Eliot's most blasphemous poem.

Ideas of childishness, linguistic degeneration, and confusion support the central theme of the degradation of essential ritual. The barely glimpsed possibility of Christian hope plunges to crazy childishness.

> *Here we go round the prickly pear*
> *Prickly pear prickly pear*
> *Here we go round the prickly pear*
> *At five o'clock in the morning.*

'Ritual,' Eliot wrote in 1923, 'consisting of a set of repeated movements, is essentially a dance.'[17] He called *The Sacred Dance* 'an excellent study of primitive religious dances', and read how these were performed round sacred trees, but also 'how often with the decay of old faiths the serious rites and pageants . . . have degenerated into the sports of children'.[18] It was ten years since Eliot had been

---

[17] 'The Beating of a Drum' (C146), p. 12.
[18] Ibid; Oesterley (n. 11 above), quoting Frazer, p. 71.

interested in *Ecstasy and Dance Hypnosis among the American Indians*, and had read in King about the lively 'moonlight dances' of Bushmen.[19] These would resurface, bereft of fun, in the childish dance of 'The Hollow Men', as later, much more lightly in the Jellicle Cats' 'Reserving their terpsichorean powers / To dance by the light of the Jellicle Moon.'

Though anthropological theories of religious degeneration were becoming outdated, Eliot was familiar with the version propounded by, for instance, Andrew Lang in his *Making of Religion*. Tylor, about whose theories Eliot had been sceptical, was particularly strong on connecting the mental processes of savages with those of children, connections emphasized by Cornford, Frazer, Oesterley, Webb, Wundt, and others. Cornford's book on some of the greatest Greek drama had as its frontispiece a Punch and Judy show. Eliot was fascinated by the idea which Cornford had put forward 'in "The Origin of Attic Comedy", [that] this [medicine-man] Doctor may be identical with the Doctor who is called in to assist Punch after he has been thrown by his horse'.

Though primitive man may be something of a mystic, the distant voices and fading star in Eliot's poem show no 'prelogical mentality' from a golden age of childlike innocence. The distant fading signals a run-down age of degenerate belief. That it had paradisal origins is a simplistic assumption. His criticism of Oesterley's work makes it clear that for Eliot a ritual's origins may be as meaningless as is its present form. He attacks Oesterley for falling into 'the common trap of interpretation' by formulating intelligible reasons for the dancing of a primitive. Eliot asserts that it is perfectly possible to claim 'that primitive man acted in a certain way and then found a reason for it'.[20]

The last section of 'The Hollow Men' in particular places its rituals in a crazy emptiness. The verve of the nursery rhyme spins us round in a sinister way, since it is disturbing to see the familiar 'mulberry bush' of the children's rhyme replaced with the arid 'prickly pear', making the rhyme like some distorted survival of a primitive chant. Eliot's substitution makes this seem an *in*fertility dance. As an American plant, 'prickly pear' also connects with his own childhood in a society whose religious values seemed atrophied.

---

[19] Cards 13 and 15 of the card index referred to above, ch. III; King, p. 109.
[20] 'The Beating of a Drum' (C146), p. 11.

Section V's voices are the most complex. First comes the choral nursery rhyme, linked typographically with the italicized passages placed against the right-hand margin to differentiate them from the rest of the text. These appear to be parts of the Lord's Prayer, but are confused by the addition of the complaint in the same typographical form, *'Life is very long'*, appropriately enough from the more primitive world of *An Outcast of the Islands*. A third voice, represented by normal type and placed in the conventional part of the page, is itself like another ritual, a fact emphasized by the incantatory effect of semantic and lexical patterning in the repeated, only slightly varied form,

> Between the *a*
> And the   *b*
> Between the *c*
> And the   *d*
> Falls the Shadow

This tripartite distinction, easy to uphold on the grounds of typography, is complicated, however, by the fact that fragments of the italicized Lord's Prayer passage find themselves brought in from the right-hand margin to form part of the body of the text when, further truncated, they make up the liturgical stutter of

> For Thine is
> Life is
> For Thine is the

before a final all-embracing italicized section, looking back in its typography, placing, and, most importantly, its rhythm, recalls the opening nursery rhyme chorus, but gives it a universal voice which seems to include all that we have heard before in what is now a ritual chant ending with an appropriately childlike sound,

> *This is the way the world ends*
> *This is the way the world ends*
> *This is the way the world ends*
> *Not with a bang but a whimper.*

This combination of various voices, including the choric, makes the fifth section distinctly dramatic in its fragmentation, and, clearly deriving from the techniques employed in *The Waste Land*, also paves the way for the fragmentary drama of *Sweeney Agonistes*. The final complexities and confusions of 'The Hollow Men' are deliberate.

Eliot's preoccupations with the 'historical sense' and cultural transmission grew from his own upbringing, but the particular stress on
language was an important theme in the anthropology he read.
Language preserved a culture's essential social values. Mallarmé's
line which Eliot quoted in 1926, discussing '*l'halluciné*' and 'un
élément *d'incantation*' in poetry,[21] 'Donner un sens plus pur aux
mots de la tribu', was reinforced by those anthropologists who found
that the fundamental mentality of the tribesmen they studied 'no
doubt . . . is transmitted socially by means of language and concepts
without which it could not be exercised. It also implies work which
has been previously accomplished, an inheritance handed down from
one generation to another.'[22] In 'The Hollow Men' we see the
corruption of such an inheritance. Shockingly juxtaposed with an
apparently childish survival of a primitive chant, the Lord's Prayer
grows increasingly fragmented, corruptly incorporating the
complaint, 'Life is very long', while being broken up and having its
utterance prevented by 'the Shadow' in the long passages which are
partly like a new ritual, usurping the old, and partly like a long gloss
which dominates the original text of the prayer. Vital patterns
become distorted and lost in the sort of fragmentation and change
with which Eliot's textual studies in a battery of languages and his
reading of works like Cornford's had made him particularly familiar.
The mixture of religious vocabulary with philosophical 'idea/reality',
and sexual vocabulary ('desire/spasm') hopelessly rearrange the
elements discovered at the roots of religion. Such confusions may
draw, too, on Eliot's fears about the barrenness of his own life and
about losing his own creativity. Middleton Murry had stated that
Eliot would never pass beyond 'the point where discrepancy ceases
between intention and achievement'.[23] The poem's 'Shadow',
though, is made widely impersonal. It falls to stifle creative enterprise,
with the result that all that remain are dying echoes

> For Thine is
> Life is
> For Thine is the

fading away to the final '*whimper*'. These echoes are the verbal

---

[21] 'Note sur Mallarmé et Poe', *Nouvelle Revue française*, 1 Nov. 1926, pp. 525–6.
[22] Lévy-Bruhl, p. 107.
[23] J. Middleton Murry, 'The Eternal Footman' (review of *Ara Vos Prec*),
Athenaeum, 20 Feb. 1920, p. 239.

equivalent of degenerating survivals of primitive rituals whose
juxtaposition with Christian liturgy recalls *The Waste Land* and
Eliot's earlier interest in a possible universality of certain forms.
The anthropological scholars' emphasis on 'Survival in Culture'
is paralleled by one of the dominant motifs of romantic poetry:
the fading, echoing voice.[24] So Wordsworthian romanticism gives
way to Tennyson's 'And answer, echoes, answer, dying, dying,
dying' and the Arnoldian 'infinitely distant land, [from which] /
Come airs, and floating echoes' ('The Buried Life'). Whether or
not we accept Eliot's note that the title of 'The Hollow Men' draws
on both Kipling's 'The Broken Men' and Morris's 'The Hollow
Land', there is little doubt that this 'valley of dying stars' is not
so far from the landscapes of Victorian romanticism. As if to
confirm this, Eliot's earlier draft of section II used the phrase 'waking
echo'.[25] In its dying echoes of both phrases and rituals, 'The Hollow
Men' again shows romanticism and anthropology going hand in
hand.

The confusion in the world of 'The Hollow Men' is furthered by
a sense of confused identity. Words come to the reader with little
traditional sense of an author. The use of nursery rhyme, prayer, and
the splitting of the poem in section V, confirm this 'authorlessness',
and the one time when the first person singular is employed, in section
II, the reader senses a fear localized in the images used, more than
in their beholder. There are, of course, precise images employed
and identifiable resonances. 'The Hollow Men' is about failures in
creation, as well as about 'the way the world ends'. So the line,
'Shape without form, shade without colour', is reminiscent of the
essential beginning when 'the earth was without form, and void,'
but what follows is not a creation and a bringing of light, but only
an impotence, 'Paralysed force, gesture without motion', bringing
a twilight. Memories of Genesis, chapter 1 are identifiable here.
Elsewhere, though, Eliot deliberately pruned away details so as
to universalize his imagery. The final version omits the 'Tartar
horsemen' who were, Perry taught him, a degenerate form of a higher
civilization; but their 'furious force' makes them inappropriate company
for those who will on no account be remembered as 'lost /

[24] Tylor. 'Survival in Culture' is the title of chapters III and IV (vol. i).
[25] 'An early version' (Eliot's note) of 'Doris's Dream Songs' which would become
parts of I–III of 'The Hollow Men' includes this phrase in section II (King's College
Library).

Violent souls'.[26] Beyond a certain stage it becomes pointless to indicate what went into the landscape of 'The Hollow Men'. Suffice it to say that the desert landscape was no doubt encouraged by his enthusiasm in 1922 and 1923 for the work of an author whom he placed with James and Conrad—C. M. Doughty, whose 'great work *Travels in Arabia Deserts*' he recommended, mentioning also that a good essay on Doughty's prose, including quotations from it, was contained in Middleton Murry's recent book *Countries of the Mind*.[27] The landscape of 'The Hollow Men' is very much a country of the mind, drawing on many of the ideas found in the pages of the anthropologists whom he catalogued as folklore experts in 1924. Almost any of these writers might have supplied, for instance, details of the importance of the wind in primitive supernatural beliefs, but the idea of 'the wind's singing' is also a prime romantic conceit, as seen in the 'gentle breeze' which wafts through *The Prelude*, or in the notion of the Aeolian harp. The latter occurs as a trick used to fool the natives in Stevenson's 'The Beach of Falesá', where 'the sound of singing in the wind' caused by an Aeolian harp is mistaken for spirit voices near a mock-temple of imitation gods, 'idols or scarecrows, or what not'.[28] The 'voices . . . in the wind's singing' and the 'dried voices . . . quiet and meaningless / As wind in dry grass' are again transformations of romanticism linked with anthropology. Other themes, too, such as boredom, the wish to present poetry in an 'authorless' way, often through impersonal ritual, and a concern with the degeneration of culture, grew from *The Waste Land* and are shared with *Sweeney Agonistes*. 'The Hollow Men' develops simpler diction.

The move to the stage was a logical one for Eliot, so many of whose poems have dramatic qualities. In 1920, considering 'the impotence of contemporary drama', he had concluded that 'The natural evolution, for us, would be to proceed in the direction indicated by Browning; to distil the dramatic essences, if we can, and infuse them into some other liquor.' This *The Waste Land* did, but when Eliot writes elsewhere that any modern poet who applied himself to the drama would be an extremely conscious poet, using

[26] W. J. Perry, *The Growth of Civilization* (London: Methuen [1924]), p. 148; Wundt, p. 152.

[27] 'Contemporary English Prose', *Vanity Fair*, July 1923, p. 51.

[28] R. L. Stevenson, *Dr Jekyll and Mr Hyde and Other Stories*, ed. J. Calder (Harmondsworth: Penguin, 1979), pp. 150 and 152 ('The Beach of Falesá').

the historical imagination, it is clear that around the time of *The Waste Land* he was also considering writing plays.[29] If *Le Sacre du Printemps* was important for the poem, significantly in 1924 he was thinking of a new 'strict form' of drama, akin to the Russian Ballet.[30]

His 'Fragments of an Aristophanic Melodrama' began as *The Waste Land* reached conclusion. Around 1920 when Eliot, with anthropological ideas in the forefront of his mind, had attacked Gilbert Murray's translations of Euripides, Pound tried to persuade him to translate Aeschylus' *Agamemnon*, but Eliot 'sat on it for eight months or some longer period'. Pound then made his own attempt, 'having the watchman talk nigger', and the project was still on the go in December 1921.[31] In early 1922 Eliot told Pound that he was trying to read Aristophanes, apparently on Pound's recommendation.[32] Pound replied that Aristophanes might be depressing, and that to raise the heavy phallus to its desired height Dixie folk melodies might be more effective.[33] This reveals a mutual awareness of theories of the origin of Attic comedy in fertility ritual. It also shows Pound relating such ritual (and Greek drama) to jazz. He was not the first to do this. Popular American poetry of the period encouraged not only 'the coming together of East and West', but also Vachel Lindsay's attempts 'to carry . . . vaudeville form back towards the old Greek precedent of the half-chanted lyric'. Harriet Monroe looked to 'the wonderful song-dances of the Hopis and other of our aboriginal tribes' to 'revitalize' literature.[34] One of the best known of Lindsay's pieces of 'Higher Vaudeville' was 'The Congo', which linked savage chants and jazz rhythms in a rather bathetically facile way that just misses being electrifying.

> BOOM, steal the pygmies,
> BOOM, kill the Arabs,
> BOOM, kill the white men,
> HOO, HOO, HOO.

---

[29] 'The Poetic Drama', *Athenaeum*, 14 May 1920, p. 635.
[30] 'Four Elizabethan Dramatists', *Criterion*, Feb. 1924, p. 119.
[31] Pound, *Guide to Kulchur* (1938; rpt. London: Peter Owen, 1966), pp. 92–3; Pound (1950), p. 170, Pound to Eliot, Paris [24 Dec.] 1921.
[32] Pound (1950), p. 171, Eliot to Pound, London [?Jan. 1922].
[33] Pound to Eliot. Paris [?Jan. 1922] (Houghton).
[34] Harriet Monroe, 'Introduction' to Vachel Lindsay, *The Congo and Other Poems* (New York: Macmillan, 1914), pp. v, vii (quoting Lindsay), and ix.

Listen to the yell of Leopold's ghost
Burning in Hell for his hand-maimed host.
Hear how the demons chuckle and yell
Cutting his hands off, down in Hell.[35]

Though Eliot did not like Lindsay's work, he did think that it might have paved the way for better things.[36] One of these, surely, is *Sweeney Agonistes* where the worlds of jazz, primitive ritual, and Greek drama again combine in an anthropological perspective.

If 'The Hollow Men' presents the degeneration of ritual, then *Sweeney Agonistes* while remaining aware of the decline attempts a reconstruction, but that reconstruction, like *The Waste Land*, sees humanity as trapped in the torturing cycle of sexual fertility. According to Moody, this play was in draft form by April 1923.[37] But Eliot continued to work on it, and it was not published until 1926-7. This period is, if we except the years at Harvard, the period in which his interest in anthropology was strongest. 'The Beating of a Drum', whose very title relates to Eliot's ideas in *Sweeney Agonistes*, appeared in October 1923, listing Cornford and a host of anthropological writers. Earlier that year, at around the time of the drafting of the play, Eliot had invited Jane Harrison to contribute to the *Criterion*.[38] In 1924 he was describing Lewis's anthropologically influenced 'Mr Zagreus and the Split-Man' as 'a masterpiece'; the following year Eliot and Lewis discussed Perry and Harrison. In August 1923 Eliot had written to Alfred Kreymborg (who published 'Portrait of a Lady' in 1915), responding enthusiastically to Kreymborg's *Puppet Plays*, and asking for advice on manufacturing of puppets. He told Kreymborg that he was planning puppet plays of his own.[39]

Already we have seen Eliot's interest in the connection between the medicine men and Punch and Judy, as suggested by Cornford in *The Origin of Attic Comedy*. Eliot's work was never performed by puppets, but Cornford's influence remained central, and a 1933 letter shows that other ideas expressed in 'The Beating of a Drum'

[35] Mark Harris, ed., *Selected Poems of Vachel Lindsay* (New York: Collier Books, 1967), p. 48.
[36] Letter to Michael Roberts, London, 19 July 1935 (Berg).
[37] Moody, p. 114.
[38] Harrison's reply of 2 June 1923 is preserved in bMS Am 1432, Houghton.
[39] 'a masterpiece'—Eliot to Lewis, 13 Mar. 1924—Lewis, *Letters*, p. 139 n. 2; Kreymborg's letter of 14 Jan. 1924 is in Houghton.

played their part in Eliot's concept of his play.

In 'Sweeney Agonistes' the action should be stylized as in the Noh drama —
see Ezra Pound's book and Yeats' preface and notes to 'The Hawk's Well.'
Characters *ought* to wear masks; the ones wearing old masks ought to give
the impression of being young persons (as actors) and vice versa. Diction
should not have too much expression. I had intended the whole play to be
accompanied by light drum taps to accentuate the beats (esp. the chorus,
which ought to have a noise like a street drill). The characters should be
in a shabby flat, seated at a refectory table, facing the audience; Sweeney
in the middle with a chafing dish, scrambling eggs. (See 'you see this egg.')
(See also F. M. Cornford: 'Origins of Attic Comedy', which is important
to read before you do the play.)[40]

Several points are significant about this passage. First, Eliot's stress
on a ritualistic presentation. He was familiar with Pound's writing
on Japanese drama, and had been impressed by the first performance
of *At the Hawk's Well* when he had seen a well-known Japanese
dancer take the role of the hawk. Moreover, he had emphasized
to Kreymborg his wish to preordain every move and gesture and
grouping.[41] Such a notion of 'preordaining' strongly suggests Jane
Harrison's concept of the *dromenon*—the rite *pre-done* and *re-done*.
The drum taps seek to restore the drum whose loss in modern drama
is stated at the end of 'The Beating of a Drum', while the striking
suggestion that the choral passages 'to have a noise like a street drill'
corresponds to Eliot's praise of *Le Sacre du Printemps* for its uniting
the noises of primitive and metropolitan life. Like *The Waste Land*,
Sweeney Agonistes explicitly unites the savage and the city. Events
in a modern flat follow the *schema* according to which Cornford
analyses Aristophanic comedy, relating it to fertility ritual. Cornford's
analysis divided the plays initially into two sections.

Of these two parts, the first normally consists of the *Prologue*, or exposition
scenes; the Entrance of the Chorus (*Parodos*): and what is now generally
called the *Agon*, a fierce 'contest' between the representatives of two parties
or principles, which are in effect the hero and villain of the whole piece.[42]

The other main part of the plays was the *Parabasis*, normally split

---

[40] 1933 letter to Hallie Flanagan quoted in Johannes Fabricius, *The Unconscious
and Mr Eliot* (Copenhagen: Nyt Nordisk Forlag Arnold Busck, 1967), pp. 123-4.
[41] 'Ezra Pound', *New English Weekly*, 31 Oct. 1946, p. 27; Eliot's phrase is
quoted in the Kreymborg letter cited in n. 39 above.
[42] Cornford, p. 2.

into two parts and accompanied by one or more *Scenes* and a *Chorikon*, before a finale often involving a feast or rejuvenation, and leading to a final *Exodos*. As well as outlining this structure in the body of his text, Cornford supplied summary outlines of the plays in a 'Synopsis of the Extant Plays'. *The Birds*, for instance, follows the pattern

PROLOGUE
PARODOS
AGON
PARABASIS
SCENE
PARABASIS II
Two Messenger announcements
CHORIKON
CHORIKON
SCENE
Another messenger announcement
EXODOS[43]

Cornford's synopses emphasize such ritual elements as a 'FIGHT', 'COOKING', leading to a revival of life, the 'FEAST' and the 'SACRIFICE', and, as in *The Birds*, the 'Invocation of the Muse'. Certain parts of the play, frequently the *Parabasis*, would contain material relating to contemporary events, originally theological, but later political. Interruptions of the action by figures of the *Alazon* (Impostor) type, marked 'AL.' by Cornford, were also common. Those buffoon figures (related by Cornford to incidents in Punch and Judy plays) were often figures from contemporary life who are eventually driven from the stage.

Eliot's pencilled plan for *Sweeney Agonistes* follows remarkably closely the format of Cornford's 'Synopses.'[44] The numbers on the left are proposed numbers of lines.

150	*Prologue.*	Doris & Dusty.
	Parodos.	Arrival of Wauchope, Horsfall, Klipstein, Krumpacker, Servants,
300		Snow. FIGHT between Servants & Snow. Sweeney arrives.

[43] Ibid., pp. 232–3.
[44] Draft Synopsis of *Sweeney Agonistes* (King's College Library).

250	*Agon.*	Sweeney monologues.
100	*Parabasis.*	Chorus 'The Terrors of the Night' anapests — pnigos.
350	*Scene.*	Entrance of Mrs Porter. Debate with Sweeney. Murder of Mrs Porter. Interruption by A.L. ~~Boy Scouts~~ Old Clothes man A.L. Dustman. A.L. Tenant Below. A.L. Pereira
100	*Parabasis II*	Theological discussion.[P. 124] Contemporary Politics Ode Antode epyrheme,
50	*Chorikon.*	Half-choruses. αἰσχρολογίαι Invocation of the Muse

1300

[second page]

1300

| 250 | *Scene.* | Sweeney begins scrambling eggs. Distribution of eggs. Return of Mrs Porter |
| 50 | *Exodus.* | |

1600

Eliot is here making use of the most modern anthropologically based classical scholarship in order to construct his own primitive play. The artist is being at once more *primitive* as well as more civilized, than his contemporaries. The figure written to the right of '*Parabasis II*' appears to be 'P. 124'. In *The Origin of Attic Comedy*, page 124 is the beginning of the section entitled 'The Second Part of the Parabasis'. As the rite is planned, the London bank clerk adds up correctly the figures in the left-hand column.

One of Cornford's most important implications for Eliot was his redefinition of what 'comedy' meant in terms of Greek and more primitive drama. The concluding chapter of his *Origin* was entitled 'Comedy and Tragedy', and stressed 'the derivation of Comedy and Tragedy from similar ritual performances', differing only 'as emphasis were thrown on the conflict and death of the hero, or on the joyful resurrection and marriage that followed'.[45] For Eliot in 1923 all

---

[45] Cornford, pp. 215 and 212.

primitive or savage art contained comic and tragic elements, comedy and tragedy being not just late but maybe also 'impermanent intellectual abstractions'. Eliot stressed that his own opinion was that once these abstractions had developed in the course of several generations within a civilization they required either renewal or else replacement.[46] A concern to return to such primitive roots no doubt explains why in his draft synopsis he originally used the word 'Melocomic' of what would later become his 'Fragments of an Aristophanic Melodrama'. Melodrama, perched on the uneasy edge between tragedy and the horribly grotesque, points also to the music hall, where Eliot had found those last vestiges of English myth. Now, through anthropology, he renewed them.

Eliot's synopsis included the murder of Mrs Porter. Cornford's *'Resurrection Motive'* showed that the aspect of physical violence in Old Comedy extended beyond 'horseplay' to the killing of a principal adversary, sometimes a villain. Yet sometimes, as in the English Mummers' Play, 'it is the representative of the good principle that is killed by the evil, and afterwards brought back to life'. It is not hard to fit Mrs Porter's death and resurrection into Cornford's schemes. Nor is the final planned dispensation of eggs out of place in a basic ritual scheme in which 'the cooking and eating of a Feast' are 'canonical', part of a 'sacramental meal . . . through which the God passes to his resurrection'.[47] In the first London stage production, Sweeney pursued Doris with a razor, and her screams were heard offstage.[48] Eliot, who invited his friends to this performance, who attended some of the rehearsals, and who expressed his full confidence in the director Rupert Doone, continued to follow in his text the original impetus which Cornford had given him.[49] Eliot's irony produced

SWEENEY. I'll be the cannibal.
DORIS.   I'll be the missionary.
   I'll convert you!
SWEENEY.   I'll convert *you!*
   Into a stew.
   A nice little, white little, missionary stew.

[46] 'The Beating of a Drum' (C146), p. 11.
[47] Cornford, pp. 83–4 and 99–100.
[48] Notes from p. 11 of a 12-page typescript of *Sweeney Agonistes* (Berg).
[49] Letter to Rupert Doone, 9 Nov. 1934 (Berg).

I have already cited in relation to 'The Hollow Men' Cornford's idea of the degeneration of ritual into debased formats. Concerning masks in the drama, he writes how at first the characters who wore them took part in 'A certain unvarying ritual action, . . . a religious mystery' which became 'grotesque'.[50] All Eliot's characters in *Sweeney Agonistes* are grotesque. Their actions are part of his reconstruction and interpretation of a once potent ritual, part of his own '*Vitalizing of the Classics*'.[51] Yet the question arises whether the ritual performed in the London flat has any validity. Can the fundamentally religious ritual be the new foundation of a piece of modern art, to be evaluated according to purely aesthetic criteria? Such questions preoccupied Eliot at the time. He wondered if art could continue to exist or be justified if it left behind its primitive purposes with the result that aesthetic objects became *direct* objects of attention.[52] *Sweeney Agonistes*, as much as the later prose of Arnold, is an attempt to come to terms with this situation and to react against it. Eliot's solution was to attempt to revive what anthropology had revealed to him as the very oldest form of ritual and express it through the phenomena of stylized contemporary life, uniting the savage and the city. That he thought this possible is suggested by his comments on Frazer whom he saw not as an investigator of a remote and hence irrelevant past, but as someone whose researches are like Freud's, of apparently universal application, applying not to a particular historical period but to 'the soul'. Frazer, through setting out to study an essential ritual of the primitive dead, had done what Eliot hoped modern literature, including *Sweeney Agonistes* would do. He had become a 'vitalizing' force. Compared with Freud's, Frazer's work was 'of perhaps greater permanence', because for Eliot it was neither a collection of data nor a theory.

The absence of speculation is a conscious and deliberate scrupulousness, a positive point of view. And it is just that: a point of view, a vision, put forward through a fine prose style, that gives the work of Frazer a position above that of other scholars of equal erudition and perhaps greater ingenuity, and which gives him an inevitable and growing influence over the contemporary mind.[53]

---

[50] Cornford, p. 202.   [51] 'A Prediction' (C153), p. 29.
[52] Review of W. J. Perry, *The Growth of Civilization and The Origin of Magic and Religion*, *Criterion*, July 1924, p. 491.
[53] 'A Prediction' (C153), p. 29.

Along with other scholars Frazer had produced a new view of classical studies which would, in Eliot's view, have a deep effect on future literature. Eliot had long been accumulating the anthropological knowledge which would affect the literature of his own future. As we have seen, it was not just the anthropology which he read at the time of composition which played its part in the great works of the early 1920s. In 1926, for instance, he returned to Westermarck (whom he had read eleven years before) attacking his ideas that the origin and development of the ideas of good and evil were caused merely by economic, genetic, and hygienic factors.[54] Yet if Eliot continued to remember his anthropological reading, this does not mean that his view of it remained unchanged. In 1913 he had attacked Frazer for attempting to explain the savage using the ideas of modern life, offering interpretation rather than fact. In 1924, though, Eliot has come to perceive *The Golden Bough* as a 'stupendous compendium of human superstition and folly', seeing in it increasingly less 'interpretation', so that it has become 'a statement of fact' which is not involved in the maintenance or fall of any theory of Frazer's.[55] Yet to present the deeds of savages as 'folly' is a crucial act of theoretical interpretation which undermines all claims to be presenting mere 'fact'. Eliot in rereading Frazer omits to apply his own frequently used 'difference between an interpretation and a fact'; his blindness to the Frazerian interpretation which equates savage fact with folly comes about surely because Eliot himself sees the two as one.[56] He was quite prepared to argue the case that savage practices might be integral to a culture: 'We may not like the notion of cannibalism or head-hunting, but that it formed part of a distinct and tenable form of culture in Melanesia is indisputable.'[57] But this does not mean that he is advocating head-hunting. Here, as more explicitly in 'Marie Lloyd', he uses the word 'culture' in an anthropological sense which stresses its general inclusiveness. It would be absurd for Eliot to advocate that head-hunting would reinvigorate British society, just as it is absurdly funny to listen to the description of Sweeney's 'missionary stew'. Yet when such an absurdity threatens to become literal, its 'folly' becomes horrific. Compelled to re-enact

---

[54] *Lectures on the Metaphysical Poetry of the Seventeenth Century*, The Clark Lectures, 1926 (TS in Houghton), Lecture VIII, p. 1.

[55] 'A Prediction' (C153), p. 29.

[56] 'The Beating of a Drum' (C146), p. 11.

[57] 'A Commentary', *Criterion*, Jan. 1925, p. 163.

rituals which appear only as stupid, trapped within the cyclic world of '*the love of created beings*', Eliot's characters lead 'preordained' lives of deep horror. A passage from *Notes Towards the Definition of Culture*, relying on ideas from 1913, hints how the poet was using his anthropological reading when discussing in an anthropological context the difference between imaginative understanding and lived experience.

Understanding involves an area more extensive than that of which one can be conscious; one cannot be outside and inside at the same time . . . The man who, in order to understand the inner world of a cannibal tribe, has partaken of the practice of cannibalism, has probably gone too far: he can never quite be one of his own folk again.*

** Joseph Conrad's *Heart of Darkness* gives a hint of something similar.[58]*

Nowhere in Eliot's anthropological reading had he come across an example of a student of primitive civilization who had himself turned cannibal. But his mind naturally associates anthropology with artistic works dealing with primitive life, since these are the ultimate attempts at 'imaginative understanding'. *Heart of Darkness* mattered not for giving Eliot the anthropological reader a better idea of how primitive society worked, but for showing Eliot the poet that the life of the savage and that of the modern urban clerk intersected on the deepest level, the level of 'The Horror' which was, essentially for him, the realization of evil within man. The work of art was especially fitted to reveal this, since its audience may both criticize, yet at the same time be involved in, its enactment. The ritual nature of *Sweeney Agonistes* emphasizes and encourages the notion of participation in a ritual process, through its use of those aspects of culture in which the people of the period participated most: jazz, music-hall, songs, melodrama. Eliot's use of primitive ritual to reveal the horror of Sweeney's existence relates to his connection of Conrad and James as writers who involve their audience in the apprehension of deep-seated levels of evil.

The great affair not to relate, but to make the reader supply, *cooperate*. Preface to *Turn of Screw*. How to suggest Evil.
    This not weakness. I speak of T. of S. and H. of D. together because both atempt to present Evil . . . Evil is rare, bad is common . . . Real Evil is to Bad just as Saintliness and Heroism to Decent Behaviour. Nothing

[58] *Notes*, p. 41.

frightful in Kurtz participating in native rites—or in Miles using vile language to schoolboys.[59]

These notes date from 1932–3, but they make explicit what is implicit in the use of primitivism in *Sweeney Agonistes*. Eliot's play aimed to penetrate beyond the 'frightful' to a deeper sense of universal evil—'*Any* man might do a girl in' (my italics). This is manifested through the unity of supposedly civilized life and the most elementary barbarity on a level which appears to be that of the basic 'need' which Eliot had accused the anthropologists of ignoring.[60] This need seems caused partly by Original Sin closely bound up with sexuality. Eliot was tormented, like Orestes. Locked in a marriage with a wife who showed increasing signs of mental instability and whose health constantly brought her to the point of death, exhausted in his work and fearing for his own life and sanity, Eliot's view of relations between the sexes is at its bleakest, as is shown not least in the epigraphs, which were originally further universalized by including (in the draft synopsis),

> Casey Jones was a fireman's name;
> In the red light district he won his fame.
> OLD BALLAD.

In *Sweeney Agonistes* it is not mere action which matters. As the fragments stand, Doris is not actually murdered. What matters is the effect of basic instinct which Eliot had earlier presented as the result of actual murder in a simpler urban context when he described in 'Eeldrop and Appleplex' the living death experienced by a man who has murdered his mistress. Such a man is already living in a different world. He has 'crossed the frontier'.[61] In *Sweeney Agonistes*, the crossing of the frontier is more explicit. Not only is it a move into a realm where the distinction between life and death seems, as in 'The Hollow Men', to have broken down, it is also a crossing to the world of the 'cannibal isle' which is revealed as only another version of the life left behind, and where the realization that '*Life is very long*' is not escaped from, but reinforced.

That phrase from *An Outcast of the Islands* is a reminder that *Heart of Darkness* was not the only Conrad novel in Eliot's mind during this period. Willems in Conrad's story, is another one of the

[59] Notes for English 26: Lectures (1933), VIII, p. 3 (TS, Pusey).
[60] Grover Smith (1963), p. 76.   [61] 'Eeldrop' (C40), p. 9.

clerks whose lives so fascinated Eliot. 'A confidential clerk' in Macassar, Willems finds his marriage going wrong, and himself, obsessed with an Arab woman with whom he flies to a remote tropical island, becoming a savage. In its 'threatening twilight' over a 'dead land', this work of fiction meshed with Eliot's anthropological reading to provide the means of expression for his personal agony in 'The Hollow Men', and in *The Waste Land*.[62] Few men other than Eliot have read through the lengthy catalogues of ritual in Frazer's volumes and found those volumes to be 'throbbing at a higher rate of vibration with the agony of spiritual life'.[63] Much of this agony Eliot imported from his own sufferings, though, as usual, these are transmuted through his reading. The agony of *An Outcast of the Islands* is heightened by several occasions when the meeting of eyes between those who have been in love is one of terrifying intensity. Eyes 'moving no more than the eyes of a corpse . . . follow me like a pair of jailers . . . They are big, menacing—and empty. The eyes of a savage . . . ' At the same time as this Conrad work was in his head, Eliot received anthropological proof that '*Life is very long*' for 'civilized' people and for 'savages' alike in an age of degeneration, and that if the savage revealed a level of horror hidden in city life, then similarly civilized man might corrupt the savage. A passage from Eliot's 1923 'London Letter' in appreciation of Marie Lloyd, shows several of the ideas discussed above: fatal degeneration of the necessary participation in cultural actions, the inanity of modern life, and the uniting of city and savage. Eliot describes how working men who went to the music hall to see Marie Lloyd and who joined in the chorus were performing part of the business of acting, collaborating with the artist in a manner essential in all art, most obviously (for Eliot) in dramatic art. The cinema lulls such working men into an apathy similar to that shown towards art by the bourgeoisie and upper classes, Eliot argues. Then, from discussing modern urban life, Eliot makes a remarkable leap.

In a most interesting essay in the recent volume of Essays on the Depopulation of Melanesia the great psychologist W. H. R. Rivers adduces evidence which has led him to believe that the natives of that unfortunate archipelago are dying out principally for the reason that the 'Civilization' forced

---

[62] Joseph Conrad, *An Outcast of the Islands* (1896); rpt. Harmondsworth: Penguin Books, 1975), pp. 38, 43 ('Life is very long'), 200 and 157.
[63] 'A Prediction' (C153), p. 98.

upon them has deprived them of all interest in life. They are dying from pure boredom.[64]

Eliot goes on to envisage a future in which applied science replaces each theatre by a hundred cinemas, each musical instrument by one hundred gramophones, each horse by one hundred cheap motor cars, with the result that the population of the whole civilized world speedily follows the lot of the Melanesians. The relationship between this passage and the 'There's no telephones' passage of 'Fragment of an Agon' is clear, and has often been indicated. Though Rivers gave some slight encouragement to Eliot in suggesting that even in our own society, religious changes have unforeseen and far-reaching effects parallel to those caused by the abolition of head-hunting in Melanesia, Eliot's linking of 'cannibal isle' and that 'slick place' London goes directly against the main thrust of the book which stresses 'the almost immeasurable difference between Melanesian and European cultures, and the sharpness of the line which still divides them where they come in contact'.[65] Fictional stressing of deep connections between primitive and developed man, as well as the connections between primitive ritual and modern dramatic and religious practice stressed by Cornford, Harrison, Frazer, and the others encouraged Eliot to interpret the Rivers book in a way which let him see in it a reflection of his personal crisis. In the Marie Lloyd piece there are hints at Eliot's childhood linking of 'civilized and uninteresting', as he laments the destruction of primitive South Sea life.[66] Certainly the author of 'The Man Who Was King' for whom savages were interesting, if rather silly, became the reader of the romantically tinged Frazer, but essentially Eliot's view of 'civilization' went deeper than gramophones. There was, though, a sometimes subliminally romantic tinge to much of the anthropology Eliot read, not least in Frazer. The criticism of anthropology, from Lévy-Bruhl's attack on the nineteenth century to Lévi-Strauss's critique of Lévy-Bruhl, followed by Derrida's deconstructions of Lévi-Strauss, has largely been an expulsion of perceived romanticism.[67]

[64] 'London Letter', *Dial*, Dec. 1922, p. 662; revised, *SE*, pp. 458–9.

[65] W. H. R. Rivers, ed., *Essays on the Depopulation of Melanesia* (Cambridge: Cambridge University Press, 1922), Introduction, p. xi.

[66] 'The Man Who Was King' (C4), p. 1.

[67] Claude Lévi-Strauss, *The Savage Mind* (1962; trans. Chicago: University of Chicago Press, 1966), p. 268; Jacques Derrida, 'Structure, Sign and Play in the

Sweeney Agonistes also recalls Eliot's childhood when he wrote of the 'South Pacific'. Eliot at twenty-one recalled reading *The Ebb-Tide*.[68] The theme of the importation to the South Seas of disease and decadence which opens Stevenson's novel would be the theme of *Essays on the Depopulation of Melanesia*. In *The Ebb-Tide*, also, the exiled American and two Englishmen, one of whom is a 'London clerk', initially rot on the beach with 'memoirs of the music-hall'. Later, confronted with the ambiguously dominating Attwater who kings it over the natives on his atoll, one of them 'broke into a piece of the chorus of a comic song which he must have heard twenty years before in London: meaningless gibberish that, in that hour and place', seemed hateful as a blasphemy: "Hikey, pikey, crikey, fikey, chillinga—wallaba dory." ' Other music-hall songs punctuate the text. The action takes place thousands of miles from 'the perennial roar of London', but the reader remains very much aware of London and the civilization it stands for as moral standards crumble; 'savage' and 'white man' become confused until the conduct of the whites, intent on mutual destruction, seems worse than that of 'these poor souls— and even Sally Day, the child of cannibals, in all likelihood a cannibal himself—so faithful to what they knew of good'. At one stage the story of the white man transported miraculously from Paumotuan beach to central London brings the two worlds immediately together. 'At first I was dazzled, and covered my eyes, and there didn't seem the smallest change; the roar of the Strand and the roar of the reef were like the same: hark to it now, and you can hear the cabs and buses rolling and the streets resound.'[69] The idea of the traffic as waves is a traditional romantic conceit, uniting elemental nature and man-made metropolis. Eliot's linking of noises from the Steppes with those of the London Underground is an extreme version of this conceit. In the late romanticism of Stevenson's tale of corruption in the South Seas, the islanders are not seen as virtuous noble savages, but certainly appear no worse than the white men who take advantage of them. Stevenson, like Tylor, sees his islanders as childlike. Behind *Sweeney Agonistes* lies Rivers, but Eliot's interpretation and use of Rivers owes much to the Stevensonian world of his childhood reading, where white men seek paradise with island

Discourse of the Human Sciences', *Writing and Difference* (London: Routledge and Kegan Paul, 1978), pp. 278–93.
    [68] 1909 essay on Kipling, partly reprinted in Adams (ch. IV n. 93 above), p. 160.
    [69] Stevenson, ed. Calder, pp. 178, 173, 248, 174, 218, and 181.

wives arrayed in 'the scarlet flowers of the hibiscus', only to find too often that they are condemned to a life of soul-destroying boredom where 'Night on the Beach' is followed monotonously by 'Morning on the Beach'.[70] Eliot refocuses his island world for 'the Age of Jazz', in a piece not entirely dissimilar to, but more bitter than, his earlier vaudeville experiment in 'Suite Clownesque'.

SONG BY KLIPSTEIN AND KRUMPACKER
SNOW AND SWARTS AS BEFORE

> *My little island girl*
> *My little island girl*
> *I'm going to stay with you*
> *And we won't worry what to do*
> *We won't have to catch any trains*
> *And we won't go home when it rains*
> *We'll gather hibiscus flowers*
> *For it won't be minutes but hours*
> *For it won't be hours but years*

diminuendo {
> *And the morning*
> *And the evening*
> *And noontide*
> *And night*
> *Morning*
> *Evening*
> *Noontide*
> *Night*
}

The emptiness of life reduced to mere sex and death is as awful on the island as in the flat. Rituals of divination in the flat produce a combination of these which mirrors the elements of island life. So Doris and Dusty in the flat read their cards and discover boredom in monotonously cyclical jazz rhythms and rhymes, while on the supposed 'cannibal isle' Doris discovers a similar tedium.

SWEENEY. Nothing to see but the palmtrees one way
    And the sea the other way,
    Nothing to hear but the sound of the surf.
    Nothing at all but three things
DORIS.   What things?
SWEENEY. Birth, and copulation and death.
    That's all, that's all, that's all, that's all,
    Birth, and copulation, and death.

[70] 'The Beach of Falesá', ibid., p. 109; *The Ebb-Tide*, ibid., pp. 173 and 184.

The twin emphases on the deaths of islanders and the need to repopulate produced many statistical tables in Rivers's book which stressed that

> That's all the facts when you come to brass tacks:
> Birth, and copulation, and death.

While the theme of loss of interest in life, linked to Cornford's pointing to the continual presence of phallic songs and rituals in the comedies he studied, combine in the crude punning that ensues.

DORIS. I'd be bored.
SWEENEY.   You'd be bored.
  Birth, and copulation, and death.
DORIS. I'd be bored.
SWEENEY.   You'd be bored.

Eliot deliberately presents his South Sea life in crude terms. Partly this involves his audience in material which seems at first sight the familiar stuff of the music-hall chorus. Partly, though, it shows the degeneration of a world view according to which the conventionally 'romantic' tropics, strangely remote world of the *'breadfruit, banyan, palmleaf'* have become a place

> *Where the Gauguin maids*
> *In the banyan shades*
> *Wear palmleaf drapery*
> *Under the bam*
> *Under the boo*
> *Under the bamboo tree.*

Cliché imagery reveals the banality of the vision. Life under the bamboo is just as trite as life in the city where Doris, Klipstein, and Krumpacker exchange their polite, meaningless words.

DORIS. You like London, Mr. Klipstein?
KRUMPACKER. Do we like London? do we like London!
  Do we like London!! Eh what Klip?
KLIPSTEIN. Say, Miss—er—uh—London's swell.
  We like London fine.
KRUMPACKER.   Perfectly slick.

Essentially, in the world of 'brass tacks' what connects the kingdom of the bamboo and the city of London are eggs. Sweeney scrambling eggs in his London flat is also the Sweeney of the fertility ceremonies described by Cornford, as Eliot's conjunction of notes '(See "you see this egg") (See also F. M. Cornford . . .)' reminds us.

> You see this egg
> You see this egg
> Well that's life on a crocodile isle.

The human egg as centre of fertility is at the heart of *Sweeney Agonistes* and of the rites on which it is patterned. Yet the very value of that fertility is questioned. If Eliot seeks to draw on the power of an ancient ritual, he also aims at the horrifying *frisson* obtained by re-enacting the ritual with the opposite of its traditional meaning. Speaking of the cinema in the Marie Lloyd piece, he accuses it of being 'rapid-breeding' and reducing its audience to a 'state of amorphous protoplasm'.[71] The generation of life and its source are both here seen as bad. As source of fertility, Doris dislikes the very essence of life just as much as Sweeney finds it tedious to 'drink this booze' and 'have a tune',

DORIS. I don't like eggs; I never liked eggs;
  And I don't like life on your crocodile isle.

Life is where girls are murdered 'with a gallon of lysol in a bath', and where the tedious business of words only elicits the answer that there is no point to speech. The play's action is zany and at times hilarious, full of mockery and tricks of speech. Like the quatrain poems it is funny at the same time as serious. One of Eliot's great strengths is his continuing sense of humour; even *The Waste Land* has Mr Eugenides and the pseudo-gentility of Madame Sosostris's 'dear Mrs. Equitone'. Yet through the sometimes bitter humour of *Sweeney Agonistes* the horror is apparent. At the first performance Vivienne was terrified by Sweeney, seeing him as a homicidal madman.[72] Sex in the play is grotesquely comic and horrible at the same time. The age-old idea of the sexual union as a death comes literally true in the impulse to kill which thrusts a man into a new world of horror. The murderous instincts of the sex war lead not to union but to a fragmentation which destroys the ability to comprehend life on any normal level,

> There wasn't any joint
> There wasn't any joint

[71] 'London Letter', *Dial*, Dec. 1922, p. 662.
[72] Vivienne Eliot, diary entry for 2 Oct. 1935, Bodleian.

after which horror one can only return to the banalities of modern city life which protect the individual from the terror of the deep insight. The movement of Conrad's Marlow back to Brussels and London is the movement back from the realization that death and life are one to the simple considerations of drinking, having a tune, and paying the rent, however much these may be 'nothing to me and nothing to you'. This matches darkly the closing of the gates of vision in 'Burnt Norton' where 'human kind / Cannot bear very much reality'. What seems to be about to be a tour of London given by 'a real live Britisher' —

> Sam of course is at *home* in London,
> And he's promised to show us around.

becomes at its most intense moments a journey, through distinctly modern music-hall conventions, to the most savage level of behaviour. London transforms; the 'real live Britisher' seems to know all too much about death. That original title, *Wanna Go Home, Baby?*, is casually jazzy, but the baby's essential home is not London but the primitive roots of civilization and, specifically, the womb. *Sweeney Agonistes* takes its audience back not simply to what was seen as the childhood, even babyhood of ritual and civilization, but further down the evolutionary ladder to the most primitive level of that 'amorphous protoplasm' which makes up the human egg.

And this homecoming is seen as futile, nightmarish. If 'The Hollow Men' traces the degeneration of ritual to a final babyish '*whimper*', then *Sweeney Agonistes* attempts to reconstruct the most basic rituals only to end in a bang. The play tails off into the cries and inarticulate exclamations where, escaping the situation in which 'I gotta use words' in their normal sense, language becomes a cry of despair and horror as primal as the '*Wah! Wah! Wah! wa-a-ah!*' which Oesterley had quoted from Spencer and Gillen's account of aborigines dancing around the destroyed home of a dead man.[73]

> And perhaps you're alive
> And perhaps you're dead
> Hoo ha ha
> Hoo ha ha
> Hoo
> Hoo
> Hoo

[73] Oesterley (n. 11 above), p. 216.

```
KNOCK KNOCK KNOCK
KNOCK KNOCK KNOCK
KNOCK
KNOCK
KNOCK
```

The connections in *Sweeney Agonistes* between the 'going home' to primitive life and the 'going home' to childhood relates to Eliot's linking of anthropological material with psychological studies. We have already seen him comparing Frazer with Freud. His personal troubles sent him in search of analysts for both himself and his wife, sometimes with disastrous consequences. Clearly Eliot was fascinated by the relationships posited between art and ritual, so that he stressed in 1923 that '*all* art emulates the condition of ritual. That is what it comes from and to that it must always return for nourishment.'[74] As editor of the *Criterion*, Eliot at this period often used anthropological material, himself selecting books for review and reading every word of what would appear in print.[75] So, in 1923, he printed Yeats's 'Biographical Fragment' giving the poet's dreams with learned notes on tree worship, dying gods, and falling stars (which we may be tempted to relate to 'The Hollow Men').[76] In 1924, as well as a contribution from Lévy-Bruhl on 'Primitive Mentality and Gambling', the *Criterion* carried Lewis's 'Mr. Zagreus and the Split-Man' where, as in *Tarr*, modern civilization is constantly punctured by a deeper, often clearly anthropological element in a strange world of lustrations, masked figures, and phallic hands.[77] To encourage Lewis in this direction, Eliot gave him to review in the next year *Egyptian Mummies* and *Essays on the Evolution of Man* by Elliot Smith, as well as *Medicine, Magic and Religion* by W. H. R. Rivers.[78] In 1923 he had himself mentioned Smith's work, and in 1925 he published Smith's long essay on 'The Glamour of Gold'.[79] Lewis's review of these books makes clear the nature of

---

[74] 'Marianne Moore', *Dial*, Dec. 1923, p. 597.

[75] 'A Commentary', *Criterion*, May 1927, p. 188; 'The Idea of a Literary Review', *Criterion*, Jan. 1926, p. 4; 'Bruce Lyttelton Richmond', *TLS*, 13 Jan. 1961, p. 17.

[76] W. B. Yeats, 'A Biographical Fragment', *Criterion*, July 1923, pp. 315–21.

[77] Lucien Lévy-Bruhl, 'Primitive Mentality and Gambling', *Criterion*, Feb. 1924, pp. 188–200; Wyndham Lewis, 'Mr. Zagreus and the Split-Man', ibid., pp. 124–42.

[78] Wyndham Lewis, review of G. Elliot Smith, *Essays on the Evolution of Man*, et al., *Criterion*, Jan. 1925, pp. 311–15.

[79] 'Lettre d'Angleterre', *Nouvelle Revue française*, 1 Nov. 1923, p. 622; G. Elliot Smith, 'The Glamour of Gold', *Criterion*, Apr. 1925, pp. 345–55.

their importance. He points out how for Smith, 'The similarity between the average Londoner or Berliner to-day, and the average "primitive man" . . . is insisted on', and quotes from Smith's text sentiments which reverberate throughout Eliot's own work: 'There is no innate tendency in man to be progressive.' Here were studies emphasizing the fragility of culture and producing a 'doctrine . . . infinitely more pessimistic than Darwin's'.[80] Reading Rivers and Smith a couple of years earlier, Eliot had appreciated this to the full, as his relating contemporary life to Melanesian death in 'Marie Lloyd' shows. He was only too pleased to print, and to read, other men's views which strengthened his own position. Though he quarrelled sharply with I. A. Richards's statement that Eliot had effected 'a complete severance between his poetry and *all* beliefs', the editor of the *Criterion* did not dissent from Richards's hostile analysis of Lawrence's 'reversion to primitive mentality'.[81] To make the point clear, Eliot printed Richards's criticisms immediately before Lawrence's short story, 'The Woman Who Rode Away', in the July 1925 *Criterion*.[82] As the story of a white woman who rides into the Mexican mountains to be sacrificed, naked and unprotesting, to the god of the Indians in order to maintain for them 'The mastery that man must hold, and that passes from race to race', this piece, particularly when contrasted with *Sweeney Agonistes*, makes clear the essential difference between Eliot's interest in the savage and that of Lawrence.[83] For Lawrence, the savage world was authentic and admirable. For Eliot in 1925 it was horribly inescapable. Primitive sexual rituals pursue Sweeney and Doris even into the heart of London, pursue them like Furies. What, for Lawrence, was to be embraced, was for Eliot a torment. But it was a torment which had to be seen and analysed. As Lawrence witch-doctored to the century's ills, Eliot too had his 'craving for ritual'.[84]

[80] Wyndham Lewis, 'Books of the Quarter', *Criterion*, Jan. 1925, pp. 312 and 315.

[81] I. A. Richards, 'A Background for Contemporary Poetry', *Criterion*, July 1925, p. 520. See Eloise Knapp Hay, *T. S. Eliot's Negative Way* (Cambridge, Mass: Harvard University Press, 1982), pp. 81–4.

[82] D. H. Lawrence, 'The Woman Who Rode Away I', *Criterion*, July 1925, pp. 529–42.

[83] D. H. Lawrence, 'The Woman Who Rode Away II', *Criterion*, Jan. 1926, p. 124.

[84] 'Dramatis Personae', *Criterion*, Apr. 1923, p. 306.

He was also concerned with primitive ritual as underlying developed
ritual. For the proper 'return' it seemed essential to go to the earliest
known sources. In 1925, considering a book by Cecil Sharp on the
history of dancing, Eliot made the criticism that though Sharp was
a historian, he was neither a philosopher nor an anthropologist, and
so his brief notes were not just insufficient, but actually conducive
to error. Eliot suggested that whoever studied dancing had not only
to include its highest forms—ballet and the Mass—but also primitive
dances such as the Australian ceremonies detailed by Spencer and
Gillen and Howitt (whom Eliot misnamed Hewitt). Sharp's book
is broken under Eliot's imaginative obsession. Eliot goes on to wish
for a combination of religious, anthropological, and neurological
knowledge as exemplified by Rome, Cambridge, and Harley Street.[85]
Harley Street was also the business premises of some of London's
leading psychoanalysts. Writing about the novel in 1927, Eliot stated
that 'nearly every contemporary novel known to me is either directly
affected by a study of psycho-analysis, or affected by the atmosphere
created by psycho-analysis, or inspired by a desire to escape from
psycho-analysis . . . ' Eliot emphasized the dangers of this and drew
attention to the importance of the Jamesian stress on 'the whole deep
mystery of man's soul and conscience'.[86] But there seems little doubt
that the psychoanalysts' stress on the childhood roots of adult
problems played its part in the general investigation of primitive roots
which fills *Sweeney Agonistes*.

Eliot's original design was abandoned. The carefully conceived
'return for nourishment' to the rituals exposed by Cornford exhausted
itself after two fragments. That of the 'Agon', was, as Eliot knew,
the better. In it the whole idea of the fertility rite is exploded, using
the very forms and devices of the traditional ritual. The original
conception of 'the return of Mrs. Porter' would be pointless, or at
least only a prolongation of horror, since life itself and the regener-
ation of it are seen, at bottom, as hideous. Those concepts of the
legal fiction and the ritual whose interpretation may change from age
to age are essential to the understanding of *Sweeney Agonistes* in
Eliot's own development as an artist. *Sweeney Agonistes* reperforms
a ritual, encouraging the participation of its audience—through

---

[85] 'The Ballet' (C164), p. 441.
[86] 'The Contemporary Novel', TS, pp. 2–3, and 1 (Houghton). A French translation of this piece was published as 'Le Roman anglais contemporain', *Nouvelle Revue
française*, 1 May 1927, pp. 669–75.

rhythms and songs, but distancing them in performance by its extreme stylization. Yet, the work's direction is quite the opposite of that conventionally assigned to the fertility rite. That rite is turned against itself in the way that the Lord's Prayer of 'The Hollow Men' is involved in an anti-Lord's Prayer. The 'Agon' having demonstrated with such force that the core of the rite is not glorious but hideous, there is no point in continuing with the play. This 'return to the sources' discovers either that such primitive celebrations are entirely divorced from modern sexuality, or that they always concealed a deathly horror. Since the 'history of belief and feeling' cannot be traced, it is impossible to conclude which alternative is the true one.[87] All that becomes apparent is that the ritual, once exposed as deathly, is dead. The basing of art on such a ritual is pointless, as Eliot indicated when writing of the medium about which some of his most important remarks of this period were passed. He wrote that no new ballet could be founded on a dead ritual. Revivals of quaint customs were mere amusements, 'For you cannot *revive* a ritual without reviving a faith. You can *continue* a ritual after the faith is dead—that is not a conscious, "pretty" piece of archaeology—but you cannot *revive* it.'[88] Eliot draws a distinction between fertile 'borrowing' from remote cultural sources and 'founding' art on a dead ritual. In its attempt to unite the worlds of city and savage, *Sweeney Agonistes* had been all too clearly founded on the fertility rite as presented by Cornford's investigation. This did not mean that the play itself was not exciting as drama, but it meant that for Eliot, intent on his search for a vital relation between ritual and art, this play would not suffice. All his probing to the roots of religion had discovered simply the bankruptcy of those roots, or at least their apparent irrelevance to the job of finding a meaning in modern life. The union of the savage and the city confirmed the hopelessness of both and of the means of life, and brought a sharp reaction in Eliot against the continuation of such attempts. Yet, as we have seen, the need in Eliot to relate art to ritual and to religious emotion continued unabated. The themes of savage and city in his work would continue to be important to the moulding of his thought, but would for the moment grow less prominent. His own failure in *Sweeney Agonistes* to find anything worthwhile to his personal salvation in all the elaborate investigation of primitive

---

[87] IPR, p. 11.   [88] 'The Ballet' (C164), p. 443.

cults resulted in an apparent about-turn in the direction of his thought.

Eliot's preface to his mother's *Savonarola* is a defence of non-realistic drama of the sort that *Sweeney Agonistes* embodied. But a new note has surfaced, a note which points out that while philosophical scepticism may have nourished and stimulated the mind of Greeks trained in civic religious observances, it was not nourishing to modern minds trained on nothing at all.[89] For Eliot, the search for liturgy and ritual has continued. The intense reading in mysticism emphasized by Gordon and Hay and so prominent in the 1926 Clark Lectures no doubt played its part in his 1927 conversion, but so did his allied interest in primitive civilizations. For this convinced him of how essential was religious practice to all civilizations. His exposing of the inanity and hideousness for his own time of the rites which underlay *Sweeney Agonistes* only helped confirm his choice of the only religion whose rituals seemed open and potent to him. This choice went hand in hand with his artistic development. 'The play, like a religious service, should be a stimulant to make life more tolerable and augment our ability to live . . . '[90] This conclusion is contemporary with the abandonment of *Sweeney Agonistes* and with Eliot's being received into the Church of England.

Sweeney Agonistes and 'The Hollow Men' seem blasphemous in their despair. Tormented, Eliot was searching for some sort of salvation. He could countenance violence and killing, if only they might provide a way out of the *taedium vitae* which formed his spiritual landscape. Irony was to the fore when in 1925 he wrote of the Russian Revolution, but behind it was a more important urge leading him to the poem 'Le Voyage' of his favourite Baudelaire. Eliot wrote of the huge scale of the Russian Revolution and of the huge scale of its violence, pointing out that only one result could justify such huge and terrible expense: 'Such a cataclysm is justified if it produces something really *new*: Un [*sic*] oasis d'horreur dans un désert d'ennui'.[91] Such an oasis Sweeney had discovered, when he penetrated beyond the 'material, literal-minded and visionless'.[92] This discrepancy between the illusory surface of life and what lies beyond, between Bradley's appearance and reality, is a perennial theme of Eliot's poetry. He was fascinated by those deepest drives

---

[89] Introduction to *Savonarola* (B4), p. xi.
[90] Ibid., pp. xi–xii.
[91] 'A Commentary', *Criterion*, Jan. 1925, p. 163.
[92] *Use*, p. 153.

which were as he put it in 1919 'canalizations of something . . .
simple, terrible and unknown'.[93] Those 'indestructible barriers'
which in the Clark lectures he saw as characterizing post-Cartesian
society are all too apparent in *The Waste Land* where in Bradley's
words, quoted by Eliot, 'regarded as an existence which appears in
a soul, the whole world for each is peculiar and private to that
soul'.[94] At the same time, Eliot saw men and women, whether
aware of it or not, as bound to the cycle of manifold yet meaningless
'Birth, and copulation, and death', perception of which yielded only
'The horror'. Those who comprehend this see the noblest feelings
frustrated or distorted. From 'Gerontion' to *Sweeney Agonistes*,
Eliot's work examined obsessively in various ways 'the awful separ-
ation between potential passion and any actualization possible in life'.

Whether in Kipling or Chapman or other authors, Eliot was
attracted to glimpses of a world beyond. 'Beneath all the declamations
there is another tone, and behind all the illusions there is another
vision.' These words were used specifically of Whitman's writing
about lilacs or the mocking-bird, and revealing that at such moments
his theories and beliefs dropped away like needless pretexts. So Eliot
wrote in 1926 when for him the poet whose 'horrified eyes' saw most
clearly the 'chasm between the real and the ideal' was Baudelaire.[95]
He complained in 1927 of living in an age 'so deficient in devotion
and so feeble in blasphemy'.[96] Baudelaire was much in Eliot's mind
in 1927. Marlowe in that year was for Eliot positively either a major
atheist or a major Christian, but the enchantments of Faustus were
Baudelaire's *paradis artificiels*.[97] Understandably for Eliot at this
time 'the relation of Belief to Ritual' was an 'immense problem'.[98]
In 1927, treating of the fascination with ritual on the part of a
Baudelaire who was vitally relevant to the present Eliot seeks to
rescue Baudelaire from 'the Swinburnian violet-coloured London fog
of the 'nineties'. Eliot's Baudelaire is opposed to the 'childishness'
of the decadent poets who enjoyed 'dressing up' in religious ritual
and 'a religion of Evil, or Vice, or Sin . . . But to Baudelaire, alone,

[93] 'Beyle and Balzac' (C80), p. 393.
[94] Clark Lectures, Lecture II, p. 15 (Houghton); *CPP*, p. 80 (citing Bradley in n. on l. 411).
[95] 'Whitman and Tennyson', *Nation and Athenaeum*, 18 Dec. 1926, p. 426.
[96] 'The Twelfth Century', *TLS*, 11 Aug. 1927, p. 542.
[97] 'A Study of Marlowe', *TLS*, 3 Mar. 1927, p. 140.
[98] 'Literature, Science and Dogma', *Dial*, Mar. 1927, p. 240.

these things were real.'[99] Baudelaire's perception of terrible depths
when surrounded by shallower figures sets him beside Sweeney. But
Baudelaire's perception also sets him beside Eliot himself.

The important fact about Baudelaire is that he was essentially a Christian,
born out of his due time, and a classicist, born out of his due time . . .
his tendency to 'ritual' . . . springs from no attachment to the outward
forms of Christianity, but from the instincts of a soul that was *naturaliter*
Christian.[100]

This seems to tie in closely with the Eliot who pronounced himself
in 1928 'to refute any accusation of playing 'possum . . . classicist
in literature, royalist in politics, and anglo-catholic in religion'.[101]
The similarities are made more comprehensible in the 1930 essay,
'Baudelaire', where he sees Baudelaire's Satanism as 'an attempt to
get into Christianity by the back door. Genuine blasphemy . . . as
impossible to the complete atheist as to the perfect Christian . . .
is a way of affirming belief.'[102] In his perception of the horror of
fertility cycle and sexuality, Eliot saw himself as following Baudelaire.
Sex on Sweeney's cannibal isle is said to be a business of being 'bored',
but the horror of the perception takes it out of mere 'boredom',
which was killing the Melanesians and, Eliot thought, the modern
world. So, for Eliot, Baudelaire 'was at least able to understand that
the sexual act as evil is more dignified, less boring, than as the
natural, "life-giving", cheery automatism of the modern world'.[103]
If blasphemy was 'a way of affirming belief', then it is possible to
see how Eliot's most blasphemously despairing poem, 'The Hollow
Men', led up to his conversion. Eliot's Baudelaire, too, as Moody
points out, was capable of ' "looking into the Shadow" '.[104] Though
it may never be possible fully to understand what converts a man
to a religion, Eliot's view of Baudelaire gives us one of the best guides
to his own conversion to Christianity. It came about through a
negative way more extreme than has yet been suggested. The
perception that a cruel savage cycle of fertility underlay all the
trappings of modern life only served to emphasize for Eliot that there
was a higher life, though one infinitely difficult to attain. Eliot came
to Christianity through a sophisticated decadence which he was able

[99] ' "Poet and Saint . . . " ', *Dial*, May 1927, pp. 425, 428, and 427.
[100] Ibid., p. 430.   [101] *FLA*, p. 7.   [102] *SE*, p. 421.
[103] Ibid., p. 429.   [104] Moody, p. 125, citing *FLA*, p. 71.

to see as going hand in hand with the most primitive life. In coming to terms with both he wished to pass beyond them. The Baudelaire in whose work he had found at Harvard both the city and the savage now acted as his guide; blasphemy led to and joined belief. So Boloisms accompany discussions of theology in Eliot's letters to Dobrée.[105] In few places are the results of the explorations of *Sweeney Agonistes* clearer than in his 1927 attack on Lawrence whom he sees as fatally humourless.

> Mr. Lawrence is a demoniac, a natural and unsophisticated demoniac with a gospel. When his characters make love—or perform Mr. Lawrence's equivalent for love-making—and they do nothing else—they not only lose all the amenities, refinements and graces which many centuries have built up in order to make love-making tolerable; they seem to reascend the metamorphoses of evolution, passing backward beyond ape and fish to some hideous coitus of protoplasm. This search for an explanation of the civilised by the primitive, of the advanced by the retrograde, of the surface by the 'depths' is a modern phenomenon . . . But it remains questionable whether the order of genesis, either psychological or biological, is necessarily, for the civilised man, the order of truth. Mr. Lawrence, it is true, has neither faith nor interest in the civilised man, you do not have him there; he has proceeded many paces beyond Rousseau. But even if one is not antagonised by the appalling monotony of Mr. Lawrence's theme, under all its splendid variations, one still turns away with the judgement: 'this is not *my* world, either as it is, or as I should wish it to be.'[106]

In 1927 Eliot turned sharply against the sort of '*love of created beings*' seen in *Sweeney Agonistes* and against its world. In 'Charleston, Hey! Hey!' of 1927, Eliot judged John Rodker 'up-to-the-minute, if anyone is; we feel sure that he knows all about hormones, W. H. R. Rivers, and the Mongol in our midst'. But Eliot appears to be turning his back on both savage and decadent city life. Speaking of hearing in Gertrude Stein's prose the 'hypnotic' 'rhythms' of 'the saxophone' he concluded that if this were of the future, then the future was 'of the barbarians'. But such a future, Eliot wrote, was not one in which he or his readers ought to be interested.[107] In the *Criterion* generally, after 1927 when he sent Frederic Manning the latest works of Frazer, Eliot kept anthropology away from creative writers, with the exception of Charles Madge, who was a trained

---

[105] Letters to Dobrée (Brotherton).   [106] As n. 86 above, pp. 3–4.
[107] 'Charleston, Hey! Hey!', *Nation and Athenaeum*, 29 Jan. 1927, p. 595.

sociologist.[108] Apart from pieces by Lawrence, he did not print further creative work which linked the primitive and the modern; he himself no longer reviewed anthropological books. Noticeably, Graves's 1927 review of Malinowski and Rivers is immediately followed by a review written by Eliot and dealing exclusively with Christian books.[109] In 1927 Eliot was in correspondence with William Force Stead, the Anglican priest to whom he expressed his deeply anxious spiritual commitment and who had baptized him. He admired a point which Stead had made about Polynesians and Christianity, but worried that he could not think of a Christian anthropologist.[110]

Yet, ironically, Stead's recollection of Eliot's walking in the woods, in true Frazerian style, after his baptism at Finstock in Oxfordshire on 29 June 1927, perceives just the unusual link of savage and city which Eliot might appear to have renounced: ' . . . after dinner we went for a twilight walk through Wychwood, an ancient haunted forest, "savage and enchanted". I can see Eliot pacing under the mighty oaks and pushing his way through hazel thickets in a smart suit, a bowler hat, and grey spats.'[111]

'Journey of the Magi', written in 1927, contains not only material quoted in Eliot's 1926 survey, 'Lancelot Andrewes', and recollections from Eliot's own life (some of which he catalogued when reminiscing in *The Use of Poetry and the Use of Criticism*).[112] It also looks back towards his engagement with the primitive. Like 'The Hollow Men' and parts of *The Waste Land*, this poem's setting is a desert one. The traditional landscape, however, is never mentioned, being involved indirectly through the details of 'the camels galled, sore-footed, refractory'. The poem is deliberately unconventional: no mention of gold, frankincense, and myrrh. But it is conventional in terms of Eliot's earlier poetry; though less dramatic, its conclusion is as apolcayptic as before. The reader becomes aware that, Nemi-like,

---

[108] Frederic Manning, 'A Note on Sir James Frazer', *Criterion*, Sept. 1927, pp. 197–205.

[109] Robert Graves, review of Bronislaw Malinowski, *Crime and Custom in Savage Society* and *Myth in Primitive Psychology* and W. H. R. Rivers, *Psychology and Ethnology*, *Criterion*, May 1927, pp. 247–53; T. S. Eliot, review of M. Murry, *Life of Jesus, et al.*, ibid., pp. 253–9.

[110] Letter to William Force Stead, 29 Dec. 1927 (Beinecke).

[111] William Force Stead, 'Some Personal Impressions of T. S. Eliot', *Alumnal Journal of Trinity College* [Washington], Winter 1965, p. 65.

[112] 'Lancelot Andrewes', *TLS*, 23 Sept. 1926, p. 622; *Use*, p. 148.

the birth of the new priest-king means the end of 'the old dispensation'—an entire world order—as 'this Birth was / Hard and bitter
agony for us, like Death, our death'. The 'Kingdoms' mentioned are
perfectly sensible in the poem's context, but remind readers of Eliot's
work of 'death's other Kingdom' and 'death's dream kingdom'.
Though explicitly Christian, 'Journey of the Magi' forms between
the earlier and later work a bridge over which the reader (with access
to the gospel word) may cross into the release of Christianity, the
new birth; but, denied that access, the speaker of the poem can only
seek relief in death to escape from having to return to the old way
in which he is 'no longer at ease'. This old way, 'With an alien people
clutching their gods', looks back to the savage world which Eliot
had been exploring, the world trapped in the ritual of 'birth, and
copulation, and death'. The word 'clutch' has particularly strong
sexual connotations in Eliot's work, as when Saint Narcissus writhes
'in his own clutch'.[113] Eliot had criticized Wundt for ignoring
sexuality's part in religion. By 'Journey of the Magi', however,
we have birth and death but not copulation. The reader is faced
with a renunciation both of the sexuality bound up with primitive
rites and, for the moment at least, of modern sexuality. Vickery
overemphasizes vegetation references by relating the 'temperate valley
. . . smelling of vegetation' with its 'running stream' to a particular
scene in *The Golden Bough*, and by insisting that the 'water-mill'
is that 'in which Tammuz was ground' and thus functions as 'a
reminder that death is the price of rebirth'.[114] General hints at
fertility ceremonies may be present, demonstrating another continuity
in theme between this and earlier poetry; but it is important to see
that, though its death and rebirth are also related, Christianity is
presented by Eliot as an escape from Frazerian cycles of fertility (in
the way that the Buddhist 'Shantih shantih shantih' hinted at such
an escape), not as its mere continuation.

Ariel Poems, like *Coriolan*, look back to the world of Eliot's
childhood and youth, whether to his reading of Mayne Reid's novels
or to his sailing off the coast of New England. They present us
continually with death,

> I should be glad of another death

---

[113] *CPP*, p. 605.
[114] John B. Vickery, *The Literary Impact of 'The Golden Bough'* (Princeton:
Princeton University Press, 1973), pp. 269–70.

I am dying in my own death and the deaths of those after me.

and birth or rebirth,

> Living first in the silence after the viaticum.
> 'Issues from the hand of God, the simple soul'
> Living to live in a world of time beyond me; let me
> Resign my life for this life, my speech for that unspoken,
> The awakened, lips parted, the hope, the new ships.

Copulation surfaces once, only to be transformed immediately into a more familiar element:

> Those who suffer the ecstasy of the animals, meaning
> Death

As the worlds of savage and sexuality had gone together in the earlier poetry, now, together, they seem banished. The apocalyptic Eliot gravitated in 1926 to Dekker's 'heapes of dead mens bones', and the 'very pleasing' cannibalism of Wanley's 'The Resurrection', which he quoted *in extenso* in a review.[115] Entering again, after 'The Hollow Men', his valley of the dry bones, Eliot dismembered himself with scarcely less thoroughness in the new Christian poetry of Part II of *Ash-Wednesday*, published in 1927. He tears his past life to pieces with the ferocity of the convert. He wrote to Stead in April 1928 that he felt that for reasons of compensation he required the most ascetic and violent form of discipline, and discussed having to come to terms with celibacy as a Christian.[116] *Ash-Wednesday*, dedicated, originally, 'To My Wife', deals, amongst other things, with the resignation of sex, which goes hand in hand with the abandoning of primitive ceremony, the link between such primitivism and the erotic being preserved. In July 1927 the *Criterion* published 'Land's End' by Archibald MacLeish, a poem celebrating the sea and an 'ancient people' who had lived by it with gods 'carved with the muzzles of jackals'. Looking back to this primitive folk, MacLeish uses the invocation 'O my people' (which would form the conclusion of section V of 'Ash-Wednesday'), and, after vividly conjuring up a seaboard landscape, recalls details of 'real things', including

---

[115] Review of *The Plague Pamphlets of Thomas Dekker*, ed. F. P. Wilson, *TLS*, 5 Aug. 1926, p. 522; 'Wanley and Chapman', *TLS*, 31 Dec. 1925, p. 907.
[116] Letters to Stead, 10 Apr. 1928 and 2 Dec. 1930 (Beinecke).

> love, the weed smell of it,
> Front against front, not hair blown
> Dark over eyes in a dream and the mouth gone . . .[117]

Eliot's renunciation of primitivism and sexuality recalls this phrasing while pulling away from the world which had fascinated him earlier; he sees from the fertile 'slotted window bellied like the fig's fruit' how

> The broadbacked figure drest in blue and green
> Enchanted the maytime with an antique flute.
> Blown hair is sweet, brown hair over the mouth blown,
> Lilac and brown hair;
> Distraction, music of the flute, stops and steps of the mind
>   over the third stair,
> Fading, fading; strength beyond hope and despair
> Climbing the third stair.

Ash-Wednesday's rejection of the erotic 'garden god, / Whose flute is breathless' is another move away from the preoccupation with primitive fertility and infertility in the earlier poetry with its 'Priapus in the shrubbery', and its following out of 'vegetation ceremonies'. There remain, though, links with that poetry. Simeon's 'Wait for the wind that chills towards the dead land' recalls 'The Hollow Men' with its 'wind in dry grass' and 'dead land'. 'The Hollow Men' is a hidden presence in *Ash-Wednesday*, not only with its desert landscape, but also with its 'twilight kingdom' and

> Eyes I dare not meet in dreams
> In death's dark kingdom

foreshadowing the later poem's 'dreamcrossed twilight between birth and dying'. *Ash-Wednesday*'s 'dry rock' is seen in a way which makes possible a spiritual pilgrimage apparently impossible in *The Waste Land*. Yet *Ash-Wednesday* maintains a connection with the fertility symbols of the savage world. These lead now not to mere eternal recurrence, but, redeemed, to spiritual content and the possibility of a peace beyond time in the search for the word which is both the word of God's grace, healing, and the 'cry' of the poet aspiring. The valley of the dry bones and the life-giving water of baptism are traditional Christian symbols, but Eliot remains aware of other similar more primitive traditions which lead him to invoke the Virgin in unusual terms:

[117] Archibald MacLeish, 'Land's End', *Criterion*, July 1927, pp. 14, 15, and 16.

> Sister, mother
> And spirit of the river, spirit of the sea,
> Suffer me not to be separated.

Such language glances back to nature worship. It also looks forward
to the later poetry, where the 'spirit of the river' Mississippi becomes
'a strong brown god', while the 'spirit of the sea' is that of the Atlantic
of the New England coast of Eliot's childhood. *Ash-Wednesday*, for
all its renunciation, does at times look towards the childhood of the
race, but more strongly it looks back to the poet's own childhood
with which this primitivism is associated, as Eliot looks back, in
language mixing 'Gerontion', Virgil, and a new interest in his own
childhood.

> And the lost heart stiffens and rejoices
> In the lost lilac and the lost sea voices
> And the weak spirit quickens to rebel
> For the bent golden-rod and the lost sea smell
> Quickens to recover
> The cry of quail and the whirling plover
> And the blind eye creates
> The empty forms between the ivory gates
> And smell renews the salt savour of the sandy earth

Such reminiscence of childhood presented at once as reality and
illusion came frequently to Eliot around the time of his conversion.
In the very year when, after taking British citizenship and joining
the Church of England, he wrote the famous 'Preface' to *For Lancelot
Andrewes* (dedicated to Eliot's mother), Eliot was recalling his own
childhood in his introductory remarks to *This American World*, and
his days at Fishermen's Corner in Gloucester.[118] Yet, as much as
reversion to primitivism, so settling back into childhood enjoyment
and innocence was undesirable to Eliot. Salvation comes not from
attachment to one's own past, but from controlling it in favour of
a higher goal.

> Teach us to care and not to care
> Teach us to sit still
> Even among these rocks,
> Our peace in His will

[118] 'Preface' to Mowrer (B8), pp. ix–xv; 'Publishers' Preface' to Connolly (B9),
pp. vii–viii.

Coriolan presents even more strongly the rejection of mere 'infantilism' as a trap. Combining elements from his childhood reading with his reading about recent political thought and events, Eliot shows the man-child who has not evolved a cry, and so for whom 'the lost word is lost', as trapped in a corner, paralysed for all his magnificence and that of his City.[119]

> O mother
> What shall I cry?
> We demand a committee, a representative committee,
>   a committee of investigation
> RESIGN RESIGN RESIGN

If Eliot largely renounced his interest in childhood and in anthropology during the few years that followed, he gave up neither entirely. Just as he maintained an interest in psychological theories of childhood, such as those of Iovetz-Tereshchenko, so he continued to take an interest in some of the anthropological material which came his way. In March 1928 he wrote on a theme which relates to that of *Ash-Wednesday* (Part II), hoping that Hardy's burial in more than one site would not establish a precedent and that future great men would not be 'dismembered' as if the nation were 'given over to idolatry of relics and fetishes'.[120] Eliot's linking of the savage and the city would never again be as prominent as in *Sweeney Agonistes*, but it would not be abandoned. Though less dominant it persists and is modified throughout his later work.

---

[119] For Eliot's use of Maurras see F. O. Matthiessen, *The Achievement of T. S. Eliot* (New York: Oxford University Press, 3rd edn., 1958), pp. 82–3.
[120] 'A Commentary', *Criterion*, Mar. 1928, p. 193.

VI

THE SAVAGE AND THE CITY
IN THE LATER WORK

EVOLUTION and the transmission of cultural forms, topics much discussed in Eliot's youth, continued to concern him in much of his later work, and he retained a profound distrust of the idea of evolution as progress. Anthropology reinforced, rather than dispelled, his scepticism about the notion that humanity might be improving. Sweeney still howled down Emerson.

In 1929 Eliot's important article, 'Experiment in Criticism', began by discussing the work of Herbert Read. He centred on Read's phrase 'evolution of poetry' and was particularly interested in the statement that 'the beginning of this study belongs to anthropology'. Sceptical about over-simple uses of the word 'evolution', Eliot wishes reader and writer to be aware of the detailed work of philologists and anthropologists in relation to literature. He exposes the vague way in which Read deploys scientific vocabulary, pointing out how much work had had to be done by a great many people and already popularized, before a literary critic could talk in this way. Eliot points to the work of Bastian, Tylor, Mannhardt, Durkheim, Lévy-Bruhl, Frazer, Miss Harrison, and many other anthropological writers. He is suspicious not because Read uses too much anthropological material, but because he uses too little.[1]

By this time his own anthropological reading was tailing off. It seems, for instance, that he never read Malinowski, though he did read Graves's 1927 *Criterion* review of Malinowski's work and referred in 1933 to the Trobriand Islands, site of some of that anthropologist's famous field-work.[2] Again, *contra* Lawrence, it is obvious that for Eliot the idea that modern western society should adopt savage customs is seen as ludicrous and reprehensible, since he believed that not even the lowest of civilized people could adapt themselves to such society without deteriorating and frequently also

---

[1] 'Experiment in Criticism', *Bookman*, Nov. 1929, p. 228.
[2] Graves (art. cit. ch. V n. 109 above), pp. 247–53.

corrupting the natives. But for Eliot now the choice was not between civilized and primitive society but between Christian, non-Christian, and anti-Christian orders.[3] Anthropology links with Conrad to counter the optimistic vision of a modern noble savagery, but the whole is seen in a new, explicitly Christian framework.

In the five years after 1928, the *Criterion* printed little anthropological writing. Instead, we see among contributors and books reviewed a gradual move towards some writers whom Eliot later would explicitly acknowledge as influential in the development of his own social thought — men such as Christopher Dawson and V. A. Demant. Meanwhile (as in his treatment of Westermarck in 1926),[4] Eliot began to dismiss some anthropologists who had earlier interested him. Turning from savage to city, specifically to London and its 'City Churches', he did not hope to turn again. Yet the peculiar link of primitive and sophisticated, as displayed in 'Marie Lloyd', still accompanied him. Just as he reacted strongly against Richards's suggestion that *The Waste Land* had severed poetry from belief, which carried the implication that the poem was a cultural museum, so Eliot campaigned now for the preservation of London's churches not as museum pieces, but as centres of religion. Art was not the criterion of civilization. Nor was religious development alone. Eliot drew attention to the view that cars, gramophones, and central heating are inessential to civilization, but pointed to a more subtle fallacy which equated art with civilization and so saw 'Alaskan or Solomon Island wood carvers were more "civilized" than the workmen and workwomen who turn out the bibelots of Woolworth's.' Such a process of reasoning would result in an affirmation that the times of highest religious development were thus more 'civilized' than Eliot's own.[5]

Religion had to be integrated with its surrounding society, not in a subservient position, but forming the vital heart of the culture. This Eliot had taken from his reading of anthropology, as would become clear from his later social writings, and he sought to preserve the physical and spiritual bonds of his culture in the city of London. In *The Waste Land*'s 'Notes', Eliot the admirer of Wren had drawn attention (referring to St Magnus Martyr) to what was to his mind one of the finest among Wren's interiors.[6] By 1926, however, the

[3] 'Catholicism and International Order', *Christendom*, Sept. 1933, p. 177.
[4] Clark Lectures, VIII, p. 7 (Houghton).
[5] 'Civilisation: 1928 Model', *Criterion*, Sept. 1928, p. 162.
[6] *CPP*, p. 78.

secular beauty had become distinctly secondary. Recalling St Magnus Martyr and speaking again of the need to preserve the fabric of the church, Eliot declared he would cease to appeal in the name of Christopher Wren and his school, and appeal instead 'in the name of Laud and the *beauty of holiness*'.[7] London and its churches became the initial battle in his campaign for a Christian society.

An only somewhat reformed Bolo followed him. In 1927 (when Eliot mentioned 'the Mongol in our midst') he told Bonamy Dobrée that he was preparing a small book, *The Bolovian in Our Midst*, proving that there was Bolovian blood in some of the leading figures of the day.[8] The tone here mocks both anthropology and the related theories of racial purity (as seen in the work of Frobenius) which were coming to the fore at that time. Frequent Bolovian references (drawings, some rhymes, and much Bolovian dogma) show that in the period 1927–30 Eliot found this a valuable way of letting off some of the emotional steam generated by his conversion; serious discussions of theology are juxtaposed in this correspondence with ridiculous accounts of Bolovian religious practice, one letter dated according to the day of St Cecilia, another according to that of St Gumbolumbo.[9]

The bowler-hatted (not entirely un-Bolovian) city gent photographed by his brother outside the offices of Faber and Gwyer in 1926 was determined that his new faith should be brought to the fore on his home, city ground with an emphasis on basic religious values, rather than those of sophisticated aesthetes.[10] Returning to 'Ancient Buildings: And the City Churches', he stressed the need to preserve the Church's role in society, particularly where it was threatened by urban secularization.[11] In 1928 he returned to the 'Stones of London', emphasizing that even when approaching a church building (here Westminster Abbey) as a historical monument of major symbolic value and aesthetic interest, the fact could not be ignored that, poets' corner or no, 'the Abbey was not originally designed primarily as a Pantheon, but as a Church; and a Church

---

[7] 'A Commentary', *Criterion*, Oct. 1926, pp. 628–9.

[8] Letter to Dobrée [Sept. 1927], (Brotherton).

[9] Correspondence with Dobrée (Brotherton).

[10] See Bernard Bergonzi *T. S. Eliot* (2nd edn., London: Macmillan, 1978), dust jacket.

[11] 'A Commentary', *Criterion*, Jan. 1927, pp. 4–5; 'A Commentary', *Criterion*, May 1927, p. 190.

it remains until it is "disaffected" '.[12] Eliot was concerned with the physical conditions of city life. That same year he worried over ' "Slum Areas" ' concerned about overcrowding and the possible creation of new slums.[13] But it was the city's spiritual health which most preoccupied him, though he realized spiritual and economic were not always inseparable. In a paragraph whose subtitle, 'City, City', harks back to *The Waste Land*, he expressed astonishment that in an economic slump the City was pulling down buildings to erect ever more splendid banks. Eliot was fond of familiar details of a London which, in a sense, he had known all his life, ever since the childhood reading of Sherlock Holmes which he recalled in 1929.[14] It was his own spiritual change which made possible after the poems of the early twenties a more affectionate view of London, but we should not assume that the owner of *Down the Silver Stream of Thames* had ever been totally blind to the beauty of the city.[15] In 1929, he had again returned to a familiar theme, stressing that, while the establishment of a 'National Preservation' body was worthwhile, it was especially important that its attention be directed not merely to the countryside. 'St. Magnus Martyr was not as important as Stonehenge, but still it was not insignificant.'[16] It was useless preserving relics of the primitive past while the living spiritual heart of London perished. Eliot's concerns were to make him a natural contributor to *The Rock* where the theme of the city would again be combined with a new modification of the theme of the savage.

Meanwhile, though, his interests in much of his prose gravitated towards the city and the consideration of social order. Introducing Johnson's *London*, he lauded its poet as 'the most alien figure' writing 'in this rural, pastoral, meditative age . . . a townsman . . . with no tolerance of swains and milkmaids'.[17] With Johnsonian vigour, Eliot discussed the way to run a society; he surveyed with distanced irony 'The Literature of Fascism', also printing MacDiarmid's 'Second Hymn to Lenin'. His classical and monarchist position led him to couple admiration for urbane, urban eighteenth-century verse with

[12] 'A Commentary', *Criterion*, Jan. 1928, pp. 1 and 2.
[13] Ibid., Dec. 1927, p. 483.
[14] 'Sherlock Holmes' (C283), p. 553.
[15] Partial list of Eliot's books (*c.*1933) in Vivienne Eliot, Scrap Book (Bodleian).
[16] 'A Commentary', *Criterion*, July 1929, p. 576.
[17] 'Introductory Essay' to Johnson (B15), p. 15.

more general wishes for social order, helping to develop his interest in the ideas of Maurras. Yet *Coriolan* seems to criticize such ideas when they produce an external show ('5,800,000 rifles and carbines') which may demonstrate force and flourish, yet has no central Word behind it. *For Lancelot Andrewes* (1928) was subtitled *Essays on Style and Order*. In 1930, though, Eliot wrote to Dobrée that he thought order and authority dangerous catchwords and was horrified at the modern scorning of democracy; he worried about the way national and racial characteristics were becoming contested issues. Thinking it possible that London's streets would contain martyrs cudgelled to death, he wondered if that might be no bad thing.[18] This wish for martyrdom would be a temptation dealt with in *Murder in the Cathedral*. Eliot's vision of London as a potential savage battleground is confirmed by Spender's memory of a lunch-time conversation in London during 1930: ' . . . I asked him what future he foresaw for our civilization. "Internecine fighting . . . People killing one another in the streets . . . " '[19] Eliot's relationship with London at this time is a mixture of love and hate. Through the summer of 1929 he complained to Dobrée of roasting in the modern city's horrors (yet wondered if he would find the country any more bearable).[20] The next year he wrote to Stead of having lived almost entirely in towns and, though not lacking in a feeling for nature, tending to go back to his memories of Missouri and New England for natural images.[21] In the early 1930s, from his city desk at Faber and Faber, Eliot was busy paying tribute to other cities which had educated him in his way of life and to city writers who had aided his way of writing. Introducing Baudelaire's *Intimate Journals*, he praised the Frenchman's skill in elevating 'imagery of the sordid life of a great metropolis . . . to the *first intensity*' and creating 'a mode of release and expression for other men'.[22] In 1932 he returned to *Bubu de Montparnasse*, a novel of Parisian low life which underlay passages in his urban poetry.[23] But now Eliot sounded a new note: 'an intense pity for the humble and oppressed.'[24] To make his tributes to earlier cities complete, before returning to Cambridge and

---

[18] Letter to Dobrée, 7 Nov. 1930 (Brotherton).
[19] Stephen Spender, 'Remembering Eliot', in Tate, p. 49.
[20] Letter to Dobrée, 10 Aug. 1929 (Brotherton).
[21] Letter to Stead, 9 Aug. 1930 (Beinecke).
[22] *SE*, p. 426.   [23] Grover Smith (1974), p. 20.
[24] 'Preface' to *Bubu* (B20), p. viii.

Boston in 1932 he wrote to a correspondent in St Louis in 1930, telling him how much his poetry owed to his childhood there.[25]

If Eliot had been formed by St Louis, then we should remember that for him St Louis had been not simply a city, but also 'the beginning of the Wild West'. Though the city seems to dominate much of his thought at this time, in 1930 appeared his translation of St-John Perse's *Anabase* where Christianity and mysticism often have a distinctly tribal feeling.

Men, dusty people and folk of diverse devices, people of business and of leisure, folk of the frontiers and foreign men . . . trackers of beasts and of seasons, breakers of camp in the little dawn wind, seekers of water and watercourses over the wrinkled rind of the world, O seekers, O finders of reasons to be up and be gone . . .[26]

Such a feeling would be incorporated into Eliot's own later work.

Return to America refuelled him. Before leaving England he had been hankering after his American roots, not only in those sections of *Ash-Wednesday* which recall the New England coast, but also in his prose, invoking, for instance, his old master Josiah Royce, who was now mostly forgotten, 'but a great philosopher in his day'. He was also expanding his social thought in general considerations on how 'the word "civilisation" comes to mean more and more: because it means all the things that we have gained, and want to keep and also all the good things that we have lost, and want to regain'.[27] Such thoughts were not purely abstract, but directed towards Eliot's own values and to London. In his wish for a healthy city he was becoming more and more aware of rural virtues and leaning towards the sort of suspect organicism which would be developed with a vengeance by Leavis. Defending the family as the centre of human life and the village as the basic social unit Eliot expresses his preference for London over other cities since it remains characteristically made up of a collection of villages whose borders touch, each maintaining its own local character. But he worries grimly about 'what sort of organic city can be left . . . '.[28] Eliot's thoughts on civilization would be developed in America, where his longing for the scenes of his youth would also be satisfied and would

[25] Letter to M. W. Childs, quoted in Matthiessen, p. 186.
[26] 'From "Anabase" by St-John Perse', *Criterion*, Feb. 1928, p. 138.
[27] 'Religion and Science: A Phantom Dilemma', *Listener*, 23 Mar. 1932, p. 429.
[28] 'The Search for Moral Sanction', *Listener*, 30 Mar. 1932, p. 480.

enable him to return to London to minister, witch-doctor-like, to his civilization, to purify the dialect of his adopted tribe.

In *The Use of Poetry* we see how by a reworking of his earlier, anthropologically inspired stress on poets as connected to the primitive man, Eliot adapted nineteenth-century views of the poet's task and brought them into his own later poetry. The Norton Lectures do not simply look back sadly. They also perfect a poetic reorientation and a restatement of poetic and cultural aims which leads Eliot to fare forward.

In his introductory lecture, Eliot quotes W. P. Ker's praise of Browning and nineteenth-century poets whose ' "themes are taken from all the ages and countries; the poets are eclectic students and critics, and they are justified, as explorers are justified; they sacrifice what explorers sacrifice when they leave their native home . . . " '.[29] Doubtless, the relevance of these remarks to Eliot's view of his own time attracted him to them. His work had taken him from London to the Arunta, from Boston Brahmins to Buddhist scriptures; he too was an explorer who had left his native home. 'East Coker' states 'Old men ought to be explorers.' Later, in *The Use of Poetry*, he links in one sentence Fitzgerald and Buddhism: 'I am not a Buddhist, but some of the early Buddhist scriptures affect me as parts of the Old Testament do; I can still enjoy Fitzgerald's *Omar*, though I do not hold that rather smart and shallow view of life.'[30] This connection hints again that Fitzgerald led Eliot to the East, and that his reading in the history of religion affected his view of nineteenth-century poetry. Against Arnold's 'educator's view' of poetry, he reacted strongly. Feeling that 'Arnold's notion of "life", in his account of poetry, does not perhaps go deep enough', Eliot presented his own essential requirement of poetry: that it possess 'auditory imagination'. Once again in this phrase reaction against the nineteenth century is bound to a standpoint which relies on anthropology. The notion of 'auditory imagination' relates to the poles of city and savage. It refers not simply to the sound of a poem's words, but has wider reverberations relating to Eliot's own explorations in anthropology, and to the sort of combination of primitive ritual and civilized world explored in *The Waste Land* and *Sweeney Agonistes*:

What I call the 'auditory imagination' is the feeling for syllable and rhythm, penetrating far below the conscious levels of thought and feeling, invigorating

[29] *Use*, pp. 22–3.   [30] Ibid., p. 91.

every word; sinking to the most primitive and forgotten, returning to the origin and bringing something back, seeking the beginning and the end. It works through meanings, certainly, or not without meanings in the ordinary sense, and fuses the old and obliterated and the trite, the current, and the new and surprising, the most ancient and the most civilised mentality.[31]

Such a theory was scarcely new in Eliot's mind. It reformulates the 1918 statement that 'The artist, I believe, is more *primitive*, as well as more civilized, than his contemporaries.'[32] Ranging from London to Melanesia, Eliot's artist's vision is wider than Arnold's. It is also deeper, seeing Melanesia in London, London in Melanesia. Discovery of Conradian horror beneath polite society, rediscovery of an Original Sin which burdened savage and citizen equally, connecting them however much the latter might politely protest, left unredeemed modern humanity at a level as terrible as that of primitives trapped in their cycle of fertility rites. Christianity was a redeeming force, making possible escape from the cycle, yet it did not cut man off entirely from the primitive. On the contrary, Eliot saw his poet as able to travel deeper down into 'the most primitive and forgotten' far beyond Arnold for whom ritual meant little, the 'inner light' being all. Such descents, according to a view which Eliot found 'very interesting', allowed the poet privileged access to a primitive mentality. This attitude may seem intensely romantic, but was acceptable to Eliot because sanctified through the anthropological eye. Highly charged personal reminiscence is bound up with the anthropological perspective which gives it an impersonal respectability. After wondering why out of a lifetime 'certain images recur, charged with emotion, rather than others', Eliot sees these as symbolic,

but of what we cannot tell, for they come to represent the depths of feeling into which we cannot peer. We might just as well ask why, when we try to recall visually some period in the past, we find in our memory just the few meagre arbitrarily chosen set of snapshots that we do find there, the faded poor souvenirs of passionate moments.

Eliot's note to this reads:

In chapter xxii of *Principles of Literary Criticism* Mr. Richards discusses these matters in his own way. As evidence that there are other approaches as well, see a very interesting article *Le symbolisme et l'âme primitive*

---

[31] *Use*, pp. 118–19.   [32] 'Tarr' (C68), p. 106.

by E. Cailliet and J. A. Bédé in the *Revue de littérature comparée* for April–June 1932. The authors, who have done field-work in Madagascar, apply the theories of Lévy-Bruhl: the pre-logical mentality persists in civilised man, but becomes available only to or through the poet.[33]

These 'depths of feeling' again look back to Eliot's 1920 'canalizations of something again simple, terrible and unknown', but now are linked directly to 'l'âme primitive'.[34] Cailliet and Bédé focused attention on the Symbolist poets and Lévy-Bruhl, a connection familiar to Eliot who at Harvard had read these poets and the anthropologist. References to Bergson, Ribot, and Durkheim plunged Eliot once more into his earlier concerns, offering him academic confirmation of his own stress on the poet's combining civilized and primitive. Stressing the crucial importance of rhythm in poetry and primitive life, while emphasizing how both primitive thought and symbolist poetry appeared to move outside Lévy-Bruhl's principle of non-contradiction, the authors stated that 'Le symbolisme institue la vaste expérience de la résurrection du primitif.' Such theories must have attracted the Eliot who so often had connected modern and savage life and whose own poetry had been attacked for its apparently illogical movements from one image to the next. Cailliet and Bédé also discussed Laforgue's 'Afrique intérieure'. In Symbolist poetry and primitive life, they found a sort of rapturous vision where living and dead appeared to unite. The thoughts of Cailliet and Bédé point forward to the importance of communion with the dead in *Four Quartets*, with their intense, visionary moments; more immediately the Frenchmen's stress that, like primitive thought, 'Le symbolisme, en effet, requiert tout d'abord une détente de l'attention', is paralleled in *The Use of Poetry* by Eliot's presentation of poetic creation not as an act of concentrated attention, but as a relaxation, or removal of a normal barrier.[35] Certainly mention of Cailliet and Bédé brings back Eliot to a familiar uniting of primitive and urban as, after stating that 'I myself would like an audience which could neither read nor write', he recalls *Sweeney Agonistes* (published in book form in December 1932) and speaks of the poet as 'something of a popular entertainer . . . having a part to play in society as worthy as that of the music-hall comedian'. The tone is abject, but Eliot approaches

---

[33] *Use*, p. 148.   [34] 'Beyle and Balzac' (C80) p. 393.
[35] Émile Cailliet and Jean-Albert Bédé, 'Le symbolisme et l'âme primitive', *Revue de littérature comparée*, April–June 1932, pp. 369, 372, and 371; *Use*, p. 144.

a climax. Though it does not end 'on this sort of flourish', the book's hope for the future comes in the one 'flourish' in its dejected concluding lines. Typically, for Eliot, this hope for the future, following the French anthropologists' piece, looks towards the distant past. 'Poetry begins, I dare say, with a savage beating a drum in a jungle, and it retains that essential of percussion and rhythm; hyperbolically one might say that the poet is *older* than other human beings . . . '[36]

Having said of the poet in *The Use of Poetry* that 'I do not mean that he should meddle with the tasks of the theologian, the preacher, the economist, the sociologist or anybody else', that was just what Eliot went on to do. He felt his Christian faith impelled him to it. In addition, his setting out the poet's connection with the primitive gave him the status of a societal elder, one who, lecturing in America, would pronounce on literature and society as Arnold had done less than fifty years before him. Eliot's vision of the urbane savage was very different from Arnold's, but gave him the same privileges as his Romantic predecessor. The poet who against Romantic diction forged a new contemporary language for a poetry which had been able to deal with the modern city face to face, had, to a considerable extent, reinstated Romanticism through anthropology. The theory that the poet had access to deep and primitive levels put him awkwardly alongside the Romantic 'explorers'. Eliot, in *The Use of Poetry*, quoted Jacques Rivière: 'It is only with the advent of Romanticism that the literary act came to be conceived as a sort of raid on the absolute and its results as a revelation.' The poet who would go on in 'East Coker' to write of poetry as 'a raid on the inarticulate' now pulls back from his earlier position in his wish 'to avoid employing the terms Romanticism and Classicism', and concludes that 'we are still in the Arnold period'.[37]

The Use of Poetry and the Use of Criticism showed Eliot again considering poetry's primitive roots and to some extent consciously reorienting himself towards romanticism. In *After Strange Gods* he continued his investigation, but with an attitude far from simple romantic primitivism. He revealed that the poet's connection with and knowledge of the primitive were only beginnings and not ends in themselves. Though savage rites might be interesting and even valuable, their greatest value was that, in showing the importance

[36] *Use*, pp. 152, 154, and 155.   [37] Ibid., p. 128 and 129.

of ritual in a society, they pointed towards the place which the higher Christian rituals should have in his own society. He saw this society as increasingly menaced by industrialization which, particularly in the cities, where power was now centralized, was cutting off the population from its roots.

Eliot was more hopeful about the 're-establishment of a native culture' in Virginia than in New England, he told a Virginian audience, since 'You are farther away from New York; you have been less industrialised and less invaded by foreign races; and you have a more opulent soil.' The city, traditional centre of civilization, is here seen as its enemy: partly because too cosmopolitan, partly because too industrialized. Eliot's sympathies lie not with urban industrialism of 'half-dead mill towns of southern New Hampshire and Massachusetts' but with humanized rural landscape which 'has been moulded by numerous generations of one race, and . . . which . . . in turn has modified the race to its own character. And those New England mountains seemed to me to give evidence of a human success so meagre and transitory as to be more desperate than the desert.' That mention of the desert takes us back to the territory traversed in *The Waste Land*, 'The Hollow Men', and *Ash-Wednesday*. The poet has emerged, wishing to save the city, ultimately presenting his true saving oasis as a version of pastoral. Not, however, a personal or directly romantic pastoral. Eliot's vision, which gives him his sympathies with ' "neo-agrarians" ', however 'quixotic', is a vision to counter the modern paganism of economic determinists whose theory has become 'a god before whom we fall down and worship with all kinds of music'. He sees his own vision not as personal, but rooted in tradition. This tradition is presented through an anthropological definition.

What I mean by tradition involves all those habitual actions, habits and customs, from the most significant religious rite to our conventional way of greeting a stranger, which represent the blood kinship of 'the same people living in the same place'. It involves a good deal which can be called *taboo* . . .[38]

This 'tradition' anticipates *Notes towards the Definition of Culture*, but also recalls the Eliot who had read Marett's 'Is Taboo a Negative

---

[38] *ASG*, pp. 16, 17, and 18.

Magic?'[39] It is characteristic of Eliot to move in *After Strange Gods*
from the savage notion of *taboo*, which he sees as having decayed
in our time so that it has become 'used . . . in an exclusively
derogatory sense', to the Christian notion of 'heresy' as being vital
to the interpretation of the modern world and to the health of the
(mainly Christian-based) 'tradition'. On a small scale, this shift
mirrors that in Eliot's poetry from 'vegetation ceremonies' to the
Christian rituals of *Ash-Wednesday*. His use of 'traditional' and
'heretical' allows him to sidestep the terms 'classical' and 'romantic',
so incorporating elements otherwise labelled 'romantic' into the structure
of his tradition which had been sanctified with an anthropological
definition.[40] The process is helped by Eliot's reading of and
reference to the works of Demant and Dawson (see below).

Eliot sees himself as an orthodox writer, trying to make his
audience aware of the need to contribute to their own living tradition.
He makes explicit that he is not writing as some individual Messianic
figure propounding an individualistic gospel deriving from 'the
ineffable wisdom of primitive peoples'. In *After Strange Gods* his
own view of savage and city is defined against that of Lawrence.
Eliot was engaged in a clarification of his boundaries. Mention of
'the dark gods of Mexico' signals that we are about to move on to
the writer for whom his strongest opprobrium is reserved.[41]

Eliot had known Lawrence's work for some time, but in 1931–2
he had grown particularly interested in that writer, whose 'travels
to more primitive lands' and use of Mexican divinities in *The Plumed
Serpent* were physical embodiment of Eliot's anthropological reading
and a likely reason for that title, *After Strange Gods*.[42] In 1931
Eliot reviewed Middleton Murry's biographical study of Lawrence
and went on to read Lawrence's letters, to which he referred in
1933.[43] Between 1931 and 1933 he also read *Fantasia of the
Unconscious*.[44] Probably some personal connections both attracted
and disturbed Eliot. For instance, he had been particularly close

[39] R. R. Marett, 'Is Taboo a Negative Magic?', in Rivers, Marett, Thomas, eds.,
Anthropological Essays Presented to Edward Burnett Tylor (Oxford: Clarendon Press,
1907), pp. 219–34. For Eliot's reading of this piece see Gray, p. 140 n. 100.
[40] *ASG*, pp. 18, 20, and 21.   [41] Ibid., pp. 33 and 41.
[42] Review of *Son of Woman: The Story of D. H. Lawrence* by John Middleton
Murry, *Criterion*, July 1931, p. 772.
[43] ['English Poets as Letter Writers'], report of Eliot's Lecture, *Yale Daily News*,
24 Feb. 1933, p. 3.
[44] As note 42 above, p. 773.

to his mother. Lawrence's 'emotional dislocation of a "mother-complex"', discussed by Murry, was picked up by Eliot in 1931, shortly after he had resisted return to childhood at the end of *Ash-Wednesday*;[45] in the same year he analysed Coriolanus's infantilism and his 'Mother mother'. For Eliot and Lawrence sex was a constant source of anxiety. Most important was the fact that personal circumstances and creative needs of each man had impelled him to be, in a phrase which Eliot applies to Lawrence in *After Strange Gods* but which applies equally well to himself, a 'restless seeker for myths'.[46] The similarities between Lawrence and Eliot, combined with the differences, made Eliot's attack on Lawrence particularly strong. For the man who had written the song 'My little island girl' Lawrence's blend of the sophisticated with savage sexuality was particularly disturbing. Attacking *Lady Chatterley's Lover* (which he was later prepared to defend), Eliot wrote of a 'morbidity which makes other of his female characters bestow their favours upon savages. The author of that book seems to me to have been a very sick man indeed.' Consideration of Lawrence prepares for the climax of *After Strange Gods*. Parts of *Mornings in Mexico* had appeared in the *Criterion*, and in mentioning them, Eliot returned explicitly to the theme of city and savage.

Against the living death of modern material civilisation he spoke again and again, and even if these dead could speak, what he said is unanswerable. As a criticism of the modern world, *Fantasia of the Unconscious* is a book to keep at hand and re-read. In contrast to Nottingham, London or industrial America, his capering redskins of *Mornings in Mexico* seem to represent Life. So they do; but that is not the last word, only the first.

Eliot saw the savage here as simply a base from which to start in the critique of the modern. His own critique of Lawrence leads to his climactic denunciation of his own generation when he turns himself prophet of the most orthodox sort, quoting a long passage from Ezekiel also used in *The Waste Land*.[47] It was after this critique of Lawrence's attitude to the modern industrial civilization and the savage world, joined to his own prophetic Christian stance towards his own time, that Eliot returned to London and the writing of *The Rock*.

---

[45] Ibid., p. 770.   [46] *ASG*, p. 44.
[47] Ibid., pp. 60–2; Ezek. 13: 3 ff.; cf. *The Waste Land*, ll. 20–22.

Though lacking its crackling intensity, *The Rock* looks back towards *Sweeney Agonistes* and Eliot's earlier work. The themes of the climax of *After Strange Gods* now emerge further as we are presented with industrial London, its unemployment, and those collapsing City Churches which had previously claimed his attention. *The Rock* opens with the seasonal fertility cycle which had horrified the trapped Eliot of the earlier poetry and the dramatic fragments.

> O perpetual recurrence of determined seasons,
> O world of spring and autumn, birth and dying![48]

We move as in *The Waste Land* into the financial heart of the 'timekept City'. That City is attacked in *The Rock*, but 'this London of ours' is also celebrated as a city to be worked for and redeemed by a Christianity sometimes remarkably primitive, a pared-down Christianity among the wild men, as a London church is seen as a 'House' built of

> brick laid upon brick;
> While encompassed with enemies armed with the spears
> of mistaken ideals;[49]

The inhabitants of this city are savages without Christianity:

> Men! polish your teeth on rising and retiring;
> Women! polish your fingernails:
> You polish the tooth of the dog and the talon of the cat.[50]

Occasionally the City's dilemma seems more like the problems of the Wild West:

> Remembering the words of Nehemiah the Prophet:
> 'The trowel in hand, and the gun rather loose in the holster.'[51]

Eliot's creation story, in thin verse, presents the movement from savage religions to the proper religion of his City. Beginning with the making of the world, he guides his audience through the pageant of primitive man ('worshipping snakes or trees') with an urgency that for flickering moments recalls *The Waste Land*. Sometimes biblical chant sounds too much like adaptation from simple anthropological textbooks, but we can see at any rate the exemplification of the way

---

[48] *Rock*, p. 7.   [49] Ibid., p. 78.
[50] Ibid., pp. 41–2.   [51] Ibid., p. 39.

in which Lawrence's 'capering redskins' had to be for Eliot now 'Life' but 'not the last word, only the first'.

> And men who turned towards the light and were known of the light
> Invented the Higher Religions; and the Higher Religions were good[52]

Here, among 'Prayer wheels, worship of the dead, denial of this world, affirmation of rites with forgotten meanings', some of the stuff of Eliot's earlier poetry and his anthropological researches, there comes a visionary instant of incarnation forming a link between God and man and, in Eliot's own poetry, between *Ash-Wednesday* and *Four Quartets*: 'A moment not out of time, but in time, in what we call history; transecting, bisecting the world of time, a moment in time but not like a moment of time.'[53] Religious development is counterpointed with the development of London. Desert and city were negatively linked in the earlier poetry; they are positively linked in the play. Primitive worlds of pagan savage and Old Testament prophet must not be forgotten: they are essential to the struggle between good and evil. The chorus exhorts the audience to be aware that 'The desert is not remote in southern tropics, . . . The desert is squeezed in the tube-train next to you.'[54] To forget this brings about the destruction of the values of civilization. Eliot's apocalyptic imagination again hints at those glimpses of street violence which he had revealed to Dobrée and Spender:

> It is hard for those who live near a Police Station,
> To believe in the triumph of violence.[55]

Convinced that the danger of the end of Christian society was at hand and that this would lead to the collapse of civilization, Eliot viewed London more in terms that contemporary Christians used to view Africa. Christian work in 'places, urban and suburban, must be *missionary* work', the work of 'imperial expansion' having exported the gospel like just another sort of industrial export, leaving 'much at home unsure'.[56] Eliot continued to see the danger of London reduced to a waste land where (with a momentary glance at Gerontion's coughing goat) debris lies

> In a street of scattered brick where the goat climbs,
> Where My Word is unspoken.[57]

---

[52] Ibid., p. 49.   [53] Ibid., p. 50.   [54] Ibid., p. 9.
[55] Ibid., p. 41.   [56] Ibid., pp. 55 and 20.   [57] Ibid., p. 29.

The Rock picks up several points made in earlier *Criterion* commentaries. Thus, churches are not like theatres, nor to be valued primarily as works of art. They must contribute towards constructing genuine communities, family values generating City values.[58] The City is again a place of insect-life like, drab scenes and rats, unless there is present what Eliot had earlier described as an 'organic', a sort of tribal, community sense. To bring about this sense of community, Eliot includes historical and contemporary Londoners of various social classes. The workmen's dialogue, which he co-authored, though filled with pieces of carefully observed slang, is none the less stilted and sometimes too obviously theatrical. More interestingly, Eliot returns to employing music-hall techniques at times reminiscent of *Sweeney Agonistes* and the 'Marie Lloyd' essay which had pointed out how music hall encouraged sympathetic collaboration between performer and audience. He makes a point of mentioning in *The Rock* one of his music-hall favourites, George Robey.[59] Eliot made clear that his *Pageant Play* made no pretence of being a contribution to the dramatic literature of England. 'It is a *revue*.'[60] It was this aspect, along with the play's closeness to liturgy and ritual (two linked methods of securing a sense of community), which most excited critics. Even those hostile found the play's closeness to music hall to be its strength. One of the most carefully observed pieces of vaudeville comes in the song whose female part begins,

> When I was a delicate slip of a maid
> What could eat nothin' more than a couple o' chops . . .[61]

Pound, though hostile to Eliot's Christianity, admired the song, classing it with 'Bolo';[62] clearly Eliot revelled in this composition drawing on a pleasure in music hall which dated back to his St Louis childhood, and his youth in Boston. He was gleeful about the idea of a Faber 'Vaudeville' production in 1929 where he starred as a baritone singing Bolovian Ballads and a song about a blue baboon.[63] Yet probably *The Rock* relies on a form too close to extinction at the time for it to bring the whole review to life. The later play's music-hall element seems strongest when closest to the

---

[58] *Rock*, pp. 54, 70, and 39.   [59] Ibid., p. 25.
[60] ' "The Rock" ', *Spectator*, 8 June 1934, p. 887.   [61] *Rock*, pp. 68–9.
[62] Pound, Letter to Arnold Gingrich, 30 Jan. 1935, in Pound (1950), p. 266.
[63] Letter to Dobrée, London, 30 Dec. 1929 (Brotherton). The song is probably by Georges Fourest.

more daring *Sweeney Agonistes*, bringing to this distinctly urban form some of the ironic bite which Brecht and Weill had brought to Berlin stages in 1929. This is seen most clearly in the song of Eliot's Blackshirts, whose bitter irony should be set against remarks concerning the undesirability of too many free-thinking Jews in *After Strange Gods*. Eliot has been accused of making racist remarks, yet for him 'All men are equal before God.' He had grown accustomed, even in the work of supposedly enlightened anthropologists, to terms such as 'savage', 'lower races', and 'inferior races', so that when he speaks of 'inequality', he may well be thinking of a vertical model, though he may mean simply 'difference' when he writes that

There will probably always remain a real inequality of races, as there is always inequality of individuals. But the fundamental identity in *humanity* must always be asserted; as must the equal sanctity of moral obligation to people of every race.[64]

It is unlikely that Eliot would have taken the trouble to defend Kipling against the charge of race superiority if he had believed in it himself.[65] While publishing Dawson on 'Religion and the Totalitarian State', he selected for notice in the 1934 *Criterion* a book highlighting persecution of European Jews; he wrote to Pound speaking of his offence at Pound's antisemitic remarks; with regard to the Vichy government in 1941 he wrote in *The Christian News-Letter* of his 'greatest anxiety' at news 'that "Jews have been given a special status, based on the laws of Nuremberg, which makes their condition little better than that of bondsmen." '[66] Eliot may have experienced displeasure at racial mixing in *After Strange Gods*, which he later withdrew from print as 'a bad book', but it was no coincidence that in the same year as the performance of *The Rock* he defended Frazer, but attacked Ezra Pound's favourite, Leo Frobenius, whose anthropological doctrines of racial purity he found particularly distasteful.[67] 'How you gwine ter keep deh Possum in

---

[64] *Rock*, p. 44; 'A Commentary', *Criterion*, Jan. 1936, p. 268.

[65] 'In Praise' (C478), pp. 155–6.

[66] Christopher Dawson, 'Religion and the Totalitarian State', *Criterion*, Oct. 1934, pp. 1–16; A. W. G. R., review of Jakob Wassermann, *My Life as German and Jew* and Ernst Toller, *I was a German*, *Criterion*, Oct. 1934, pp. 168–9; letter to Pound, 13 Aug. 1954 (Beinecke); 'The Christian Education of France', *Christian Daily News-Letter*, 3 Sept. 1941, p. 1; see also Moody, pp. 370–2.

[67] Behr (ch. IV n. 68 above), p. 43; ' "The Use of Poetry" ', *New English Weekly*, 14 June 1934, p. 215.

his feedbox when I brings in deh Chinas and blackmen?' Pound wrote
to Frank Morley in 1937, but for all his growing concentration on
European civilization, Eliot saw more clearly than Pound the dangers
of Fascism.[68] Moving from the elements of primitive worship to the
rituals of Christianity and the chants of the London music hall, Eliot
tried to strengthen his city against dangers of contemporary politics.

The Rock transformed the motifs of city and savage found in
Sweeney Agonistes, but they are preserved because essential to Eliot's
thought. His interest in them is confirmed by his return to the earlier
play for the production by The Group Theatre Company in the same
year as the performing and publication of *The Rock*. *Sweeney
Agonistes* was performed by Doone's company with an appropriate
mixture of anthropologically induced masked characters in hieratic
postures and music-hall routines. At its centre, the plain-suited figure
of Sweeney looks like a city clerk, surely linked to the Eliot who
jokily signed himself F. [rancis?] X. [avier?] Sweeney in 1933 when
he talked of coming over all of a hoo-hah.[69] From the condition of
being *Agonistes*, Sweeney points forward to the later ritualistic
conception of Eliot's Christian drama.

Production and publication of *The Rock* led directly to Eliot's being
invited to write *Murder in the Cathedral*, which took further his idea
of committed Christian drama and his warnings about the dangers of
a society cut off from its religious roots. It also developed his implicit
warnings against the dangers of political extremism. Bishop Bell, who
commissioned the play, was an influential British churchman who had
consistently supported the German Confessing Church in its opposition
to National Socialism.[70] Eliot's *Murder* demonstrates his agreement
with that opposition. It is a play which, again, has not entirely abandoned the traditions of the music hall. As in *The Rock*, these are
invoked in one of the most bitter moments, with the parody-hymn
not unlike a ferocious music-hall song as the chorus of knights sings

> Are you washed in the blood of the Lamb?
>   Are you marked with the mark of the beast?
> Come down Daniel to the lions' den,
>   Come down Daniel and join in the feast.[71]

---

[68] Pound, letter to F. V. Morley, Rapallo, Feb. 1937, Pound (1950), p. 288.
[69] Letter to Pound, 5 Apr. 1933 (Beinecke).
[70] Duncan Shaw, 'The Kirk and the Hitler Régime', *Life and Work*, July 1983,
pp. 26-7.
[71] *CPP*, p. 274.

Music hall had shown Eliot that the chorus of the Greek drama in whose primitive origins he had exhibited such interest, could still work on the modern stage. There are signs at times of a reversion to a more primitive level of superstition in the play, a slipping beneath Christianity to Frazerian rituals half hinted at, but suppressed beneath normal life.

> We have seen the young man mutilated,
> The torn girl trembling by the mill-stream.
> And meanwhile we have gone on living.[72]

Horror brings Eliot's chorus a sense of unnatural reversion which earlier he had found associated with anthropological discussions and had detected behind Lawrence's 'explanation of the civilised by the primitive' sending his characters back 'to reascend the metamorphoses of evolution'.

> I have lain on the floor of the sea and breathed with the breathing
>   of the sea-anemone, swallowed with ingurgitation of the
>   sponge . . .
>
>   . . . I have seen
> Rings of light coiling downwards, descending
> To the horror of the ape.[73]

A similar horror sends back the chorus far from their supposedly civilized city into a desert past where April would be again the cruellest month, as they pray 'let the spring not come'.[74] One problem with many passages in Eliot's plays is that they send us to similar but stronger passages in the poetry. The essential thing which the plays carry on from Eliot's earlier anthropologically concerned work is a fascination with ritual. The line of dramatic form running from primitive arrow dance to Pinero is worth bearing in mind.[75] Music hall was the living form closest to the liturgical extreme. In *The Rock* Eliot included song and dance. Song persists in *Murder in the Cathedral*, but Christian liturgy takes over almost entirely. Yet Eliot seems to have been attracted even to the most unlikely sources, including even Conan Doyle's 'The Musgrave Ritual'.[76] In 1934, the year of *Murder in the Cathedral*'s commissioning, he

---

[72] *CPP*, p. 257.   [73] *CPP*, p. 270.   [74] *CPP*, p. 275.
[75] 'Introduction' to *Savonarola* (B4), p. x.
[76] See, e.g., Nathan L. Bengis, 'Conan Doyle and T. S. Eliot', *TLS*, 28 Sept. 1951, p. 613.

wrote, 'One of the reasons why the *Morte Darthur* is a permanent source of refreshment, is the degree to which the primitive "ritual" stories are and are not integrated into the narrative.' Eliot returned to familiar themes in suggesting that the morality of the *Morte Darthur* was of that primitive type which was essential, whereas modern manners were things of the surface only. 'This primitive morality was refined by Christianity; but the passing of Christianity has left only the refinement without the morality . . . '[77] The *Morte Darthur* interpreted by Weston had contributed to *The Waste Land*. Two other works with which Eliot linked Malory's 'profound, tribal, Sophoclean morality' were to form starting points for his own plays; Aeschylus' Tragedy of the House of Atreus and Sophocles' *Oedipus at Colonus* underlie respectively *The Family Reunion* and *The Elder Statesman*. If in *Murder in the Cathedral* he had chosen a particular martyrdom which functioned as a bloody, savage ritual, then in the plays which followed, connections with primitive ritual would be clearer—if anything, as Eliot later feared when thinking of *The Family Reunion*, too clear, the primitive outline getting in the way of the Christian story. Harry experiences

> The sudden solitude in a crowded desert
> In a thick smoke, many creatures moving
> Without direction, . . .[78]

As in *The Waste Land*, spring appears 'an evil time'. Harry's return to the sources at wishwood seems very much a return to the Frazerian wood, when

> Spring is an issue of blood . . .
> Do not the ghosts of the drowned
> Return to land in the spring?[79]

Momentarily it seems Harry will escape by 'going away—to become a missionary' in 'a tropical climate' and getting to know the 'natives'.[80] This hint of physical as well as psychological confrontation with primitivism predictably gives rise to the corresponding image of city apocalypse, presented by Charles, the character with whom as Eliot told Martin Browne he most closely identified.[81]

[77] 'Le Morte' (C353), p. 278.   [78] CPP, p. 294.
[79] CPP, p. 310.   [80] CPP, p. 344.
[81] E. Martin Browne, *The Making of T. S. Eliot's Plays* (Cambridge: Cambridge University Press, 1969), p. 106 (Eliot's letter of 19 Mar. 1938).

As if the earth should open
Right to the centre, as I was about to cross Pall Mall.

Primitive terrors remain dormant, continually erupting in 'various spells and enchantments', or anxious questions, 'What ambush lies beyond the heather / And behind the Standing Stones?' The modern world is juxtaposed against and involved with a world very definitely presented in terms of primitive rite—nowhere more clearly than in the ritual expiation of the curse.[82] The laying of ghosts by moving 'Round and round the circle / Completing the charm' is like a rite rescued from 'The Hollow Men', presumably in Eliot's terms sanctified by the Christian connotations also present in the play. Yet the play's overall effect is not necessarily Christian. It simply communicates something of that 'profound, tribal' morality which Eliot associated with Greek drama and its ritual. It is questionable if this play fulfils Eliot's new vision as powerfully as those *Fragments of an Aristophanic Melodrama* fulfilled his old.

Relations between *The Family Reunion* and *Sweeney Agonistes* have often been remarked on. Carol H. Smith's fine study details thoroughly connections between Eliot's late plays and the earlier dramas and primitive motifs which underlie them.[83] Having indicated such links, I shall not discuss each one, but concentrate on particular instances when the themes of savage and city come together. This conjunction is particularly important in *The Cocktail Party*.

While Eliot might mock Pound's concerns with primitive society, it continued to interest him as is indicated by allusions in his correspondence to devil doctors, Bali, and Trobriand Islanders.[84] To a degree Eliot is just reassuring his correspondent that he has not changed absolutely. Though a churchwarden, he can still tell stories of his American youth, use four letter words, or quote Bolovian rhymes and customs.

Looking back towards his study of Cornford and Harrison, he approves of Johnson's view that for the modern the distinctions between tragedy and comedy were superficial. But Eliot emphasized that Johnson, unaware that such distinctions grew out of a difference

[82] *CPP*, pp. 345, 348, and 349.
[83] Carol H. Smith, *T. S. Eliot's Dramatic Theory and Practice* (Princeton: Princeton University Press, 1963).
[84] Letters to Pound, 1 Feb., 18 June, and Childermass 1934 (Beinecke).

in ritual, had been ignorant of their importance for the Greeks. Eliot's next sentence has a familiar ring: 'In the end, horror and laughter may be one—only when horror and laughter have become as horrible and laughable as they can be . . . '.[85] Here again is *Sweeney Agonistes*. Eliot's concern that his own urban society be well founded on a necessary myth was accentuated by the realization that this society was now annihilating other very different forms of civilization. This led him in 1935 to discuss a newly 'discovered' agricultural Papuan tribe, the description of whose civilization Eliot utilized to criticize what he saw as some of the indulgences of his own inorganic civilization during the unemployment of the thirties. The case of the Tari Furora paralleled that of the Melanesians in 'Marie Lloyd', but it is to his ideal City, and his actual urban Western civilization that Eliot relates this Papuan discovery: ' . . . if we are so helpless in the hands of our "civilization" that we admit our inability to prevent it from ruining Papuans, what hope have we of saving ourselves?' He makes it clear that he believes 'that there are higher civilizations and lower ones'; the notion of Empire is 'not necessarily an ignoble one' when it is founded on 'the notion of extending law, justice, humanity and civilization—with no other interest than glory, and no other motive than a sense of vocation'.[86] It seems to him more likely though that Empire is working in tandem with exploitation. This piece is intensely pessimistic and is a forerunner of his later 'doubt of the validity of a civilization' on the eve of the Second World War.[87] Civilization had to be more than a mere confluence of economic interests: 'And until we set in order our own crazy economic and financial systems, to say nothing of our philosophy of life, can we be sure that our helping hands to the barbarian and the savage will be any more desirable than the embrace of the leper?'[88] The Tari Furora greatly disturbed him. A year later, they were still in his mind when thinking simply of western civilization; the Papuans came into his memory only to be cursorily dismissed as 'weak inaudible voices'. Yet the mention of these voices leads Eliot on to wider questioning of what civilization is. He does not try to pretend there would be any viable alternative in a 'noble

[85] 'Shakespearian Criticism, I. From Dryden to Coleridge', in Harley Granville Barker and G. B. Harrison, eds., *A Companion to Shakespeare Studies* (Cambridge: Cambridge University Press, 1934), p. 295.
[86] 'A Commentary', *Criterion*, Oct. 1935, pp. 66–8.
[87] *ICS*, p. 82.   [88] As n. 86 above, p. 68.

savage' existence. But the realization that his own civilization is extinguishing cultures of 'savages' possessed of coherent ways of living prompts him to demand that civilization be considered in no narrow perspective. He emphasized that peace had to be preserved for the entire human race and not for particular sections of it, pointing out that the word 'civilization' could be applied to a wide diversity of values.[89] *Murder in the Cathedral* addressed such matters, posed in a different way. They would be taken up in Eliot's later prose.

Though apparently divorced from 'Cultural Progress' as related to the Basuto, which Eliot was also considering in 1936, his idea of poetic drama was part of the same concern with embodying and strengthening what he had always associated with ideas of culture and community and which his dealings with the 'lower races' had helped to teach him: the need for art linked to religious ritual as a central value summing up and sustaining the social values of a culture. He was appalled at the thought of turning into a pub at five-thirty for a quiet drink and finding that it had been converted into a Poets' Pub, 'reverberating, like an African village, with the roll of "Drake's Drum." '.[90] Yet he emphasized that, unlike some other societies, his own was vitiated by a lack of connection between ritual, drama, religion, and the centre of the culture: London. In a consideration of religious drama which clearly descends from the theorizing about anthropology and literature in 'The Beating of a Drum', and from the 1923 conviction that 'the stage—not only in its remote origins, but always—is a ritual,' he writes in favour of a closer union between ritual and metropolitan drama, seeing an essentially religious craving as latent in all serious patrons of drama, as opposed to cinema audiences seeking mere distraction. Eliot stressed that there should be no maintaining of different attitudes for cathedral drama and for West End theatre. He perceived instead a need to reintegrate both these types of drama and a parallel need to 'strive towards a reintegration of life'.[91] Clearly these ideas relate to Eliot's plays and to 'Burnt Norton' where dangers of being 'Distracted from distraction by distraction' seem connected with the spectre of a metropolis seen as the antithesis of religious values.

[89] 'A Commentary', *Criterion*, Oct. 1936, pp. 65–6.
[90] Ibid., Oct. 1937, p. 85.
[91] 'Religious Drama: Mediaeval and Modern', *University of Edinburgh Journal*, Autumn 1937, pp. 12–13.

London, however organically composed of

> Hampstead and Clerkenwell, Campden and Putney,
> Highgate, Primrose and Ludgate

leads to a 'twittering world', close to that of *The Waste Land*'s London or the 'timekept city' of *The Rock*. 'Burnt Norton' retains many links with Eliot's earlier poetry. For a moment balanced against London's distracting inanity are the equally distracting primitive rituals now seen as inane because sundered from the Word:

> The crying shadow in the funeral dance,
> The loud lament of the disconsolate chimera.

The funeral dance is part of that turning world which, like the remorselessly and vapidly revolving seasons of *The Waste Land* (or the circles of the London underground in 'Burnt Norton') threatens to trap its inhabitants in a cycle which is, eventually, empty. So timeless Love is distinguished from temporal desire which,

> Caught in the form of limitation
> Between un-being and being

reminds the reader of that moment when the shadow falls 'Between the potency / And the existence', or of Tiresias's being trapped unfulfilled in sexual desire, bearing witness to all the ritualistically patterned couplings of history, 'throbbing between two lives'.

The themes of city and savage are marginally present in 'Burnt Norton'. As the *Four Quartets* develop, their importance increases.

The idea of 'distraction', so strongly linked with London in 'Burnt Norton' was one which Eliot saw in this period as politically as well as spiritually dangerous. For him spiritual and political ideas were becoming more and more inseparable in his concern with 'culture' as a whole. His considerations of political distraction and its dangers were panoramic in scope.

The modern 'dictator,' a Hitler or Mussolini, must be thought of . . . as a highly paid *leading actor*, whose business is to divert his people (individually, from the spectacle of their own littleness as well as from more useful business) . . . The ruler as dramatic star—with illustrations from the customs of African tribes—is one subject of Mr. Lewis's attention.[92]

[92] 'The Lion and the Fox', *Twentieth-Century Verse*, Nov./Dec. 1937, p. 7.

At the same period, though, Eliot continued to worry about '*urban-ization of mind*'.[93] In 1940 he wrote of an almost sanctified rural life as something vitally different from the separately developing urban culture which so disturbed him and pleaded that agriculture should be regarded as a vocation, rather than merely an industry, though he pointed out that such a view involved the whole orientation and scheme of values of a future society.[94] In *The Idea of a Christian Society*, published just before war, Eliot saw 'Our choice' as being 'between a pagan, and necessarily stunted culture, and a religious, and necessarily imperfect culture'. Unlimited industrialism is particularly dangerous, tending 'to create bodies of men and women—of all classes—detached from tradition, alienated from religion, and susceptible to mass suggestion'. Though that last phrase seems to point to Germany and Italy, Eliot also draws attention to dangerous mass 'cults' in Britain, which 'has been highly industrialized longer than any other country'.[95]

Claiming that he is not 'presenting any idyllic picture of the rural parish', Eliot takes as his 'norm, the ideal of a small and mostly self-contained group attached to the soil . . . with a kind of unity which may be designed, but which also has to grow through generations'. He was clearly attracted to rural religious communities, such as that at Kelham where he himself participated or that seventeenth-century community of Little Gidding which he was later to celebrate. Yet he saw the danger in making these communities an ideal, as he turned an anthropological eye on Christianity and perceived that such examples seem to proffer no solution to industrial urban and suburban existence—the way most people live. Such a religious patterning of small communities reveals a 'Christendom fixed at the state of development suitable to a simple agricultural and piscatorial society', and so imperfectly suited to the more complicated organization of modern society.[96] Eliot's solution of a widespread Christian community hierarchically organized, related both to the state and individual parishes and containing intellectual leaders, owes much to Benda's notion of *clercs*, as well as to the anthropologists' stress on the connection of religion with society. As in *After Strange Gods*, he is concerned in his conclusion with a return to sources,

---

[93] 'A Commentary', *Criterion*, Oct. 1938, p. 60.
[94] 'The Church in Country Parishes', *Christian News-Letter*, 28 Aug. 1940, p. 4.
[95] *ICS*, pp. 51, 53, 52, 53.   [96] Ibid., pp. 59–60.

that is with going back to the savage and working forwards towards
his solution to the problems of modern industrial life; again such
a movement is presented in terms of a familiar encounter.

The struggle to recover the sense of relation to nature and to God, the
recognition that even the most primitive feelings should be part of our
heritage, seems to me to be the explanation and justification of the life of
D. H. Lawrence, and the excuse for his aberrations. But we need not only
to learn how to look at the world with the eyes of a Mexican Indian — and
I hardly think that Lawrence succeeded — and we certainly cannot afford
to stop there. We need to know how to see the world as the Christian Fathers
saw it; and the purpose of reascending to origins is that we should be able
to return, with greater spiritual knowledge, to our own situation.[97]

Eliot does not see primitive religion as a necessary basis for Christianity,
'I do not believe that Christianity germinates out of natural religion,
but that it is given by revelation.' Though his 'City' is put forward
as a remedy for secular, increasingly industrial-urban values, he does
conceive of it as including non-Christians, though these he hopes
would be a minority.[98]

This period of Eliot's development is often seen as one of increasing
narrowing, so that it is useful to emphasize that while it certainly
represents a concentration of energy on Christian themes, it also
represents a continuing openness to other elements and a drawing
on resources whose foundations were laid in Eliot's studies at
Harvard. Discussing 'Education in a Christian Society' in 1940, Eliot
cites eastern ideas of '*sadhu*, or *mahatma*' in a demonstration that
'In the East, and in pre-Christian Europe, the sage and the saint have
been hardly distinguishable from each other. We must recognize the
truth in both the Oriental and the Christian views.' Such an openness
would be seen increasingly in *Four Quartets*. Elsewhere, he criticizes
Mannheim's definition of charismatic education in the consideration
of the awakening of religious feelings since it seems neither to include
the whole of the education of so-called 'primitive races' any more
than of the higher races in their religious stage. Eliot instances the
co-ordination of different types of education in 'the highly organized
societies of Polynesia'. The breadth of this review, which goes on
to discuss Sir Thomas Elyot's *The Governour* and to conclude
by looking at immediate reforms applicable in a modern world of

[97] *ICS*, p. 81.
[98] 'A Sub-Pagan Society', *New English Weekly*, 14 Dec. 1939, p. 126.

'industrial exploitation . . . ' where 'local community does not exist' relates clearly to 'East Coker'.[99]

This poem shows how Eliot's increasing interest in ruralism had at its heart his earlier interest in the savage. As in *The Waste Land*, this is combined with the city. 'East Coker' is concerned with civilization at its beginning and end, as well as with purely individual values. The poem's beginning is certainly biblical, drawing on Ecclesiastes, but it also involves the modern world ('factory . . . by-pass'), and deliberately seems to draw on the biblical passage closest to Frazerian fertility rites, the dead being reborn through the seasonal cycle.

> Old stone to new building, old timber to new fires,
> Old fires to ashes, and ashes to the earth
> Which is already flesh, fur and faeces,
> Bone of man and beast, cornstalk and leaf.

Eliot celebrates English pastoral rhythms. In 'East Coker', though, we seem to watch ghosts from the past, though the countryside itself may be little changed:

>   In that open field
> If you do not come too close, if you do not come too close,
> On a summer midnight, you can hear the music
> Of the weak pipe and the little drum
> And see them dancing around the bonfire
> The association of man and woman

This vision is like that which Eliot praised in his childhood favourite, Kipling, whose *Puck of Pook's Hill* and *Rewards and Fairies* particularly 'give at once a sense of the antiquity of England, of the number of generations and peoples who have labored the soil and in turn been buried beneath it, and of the contemporaneity of the past'. 'They', one of Kipling's 'most other-worldly stories' was important for the opening of 'Burnt Norton'.[100] Kipling's historical and contemporary tales of the English countryside, stories such as 'An Habitation Enforced', 'My Son's Wife', and 'The Wish House', are important to the start of 'East Coker'. The earlier Kipling brought about an 'introduction of India and the Colonies into the sphere of consciousness

---

[99] 'Education in a Christian Society', *Christian News-Letter*, 13 Mar. 1940, pp. 2–3.
[100] 'In Praise' (C478), pp. 156, 155.

of the inhabitants of the London suburb'; the later was 'discovering
and reclaiming a lost inheritance'.[101] Both were crucial to Eliot.
When he maintains the old spellings, 'daunsinge' and 'matrimonie',
Eliot links himself with his East Coker ancestor, Thomas Elyot.[102]
The poet chose burial under East Coker earth, completing a circle
of ends and beginnings. The attraction of the stately dance in 'East
Coker' is strong.

> The association of man and woman
> In daunsinge, signifying matrimonie—
> A dignified and commodious sacrament.
> Two and two, necessarye coniunction,
> Holding eche other by the hand or the arm
> Whiche betokeneth concorde.

This is, however, mixed in with a more primitive element relating
rustic to savage through fertility rituals. Eliot changed 'May midnight',
(recalling Gerontion's 'depraved May' and ceremonies of spring)
to 'summer midnight', maintaining a primitive connection with
'Midsummer Fires', ceremonies copiously described by Frazer, while
looking back, too, to the Hollow Men's circular dance.[103] As in his
remarks on Lawrence in *After Strange Gods* and *The Idea of a
Christian Society*, Eliot indicates the importance of contact with the
savage as a beginning, while stressing its limitations. He makes
apparent both his sympathies with rural life and the fact that mere
retreat to the tradition of the countryside is not enough. He has
respect for the almost tribal traditional wisdom of those whom he
later called the 'quiet-voiced elders' and their 'dead secrets', but
in the end these are to be renounced or incorporated in a higher
Christian scheme and vision. Though attractive and lyrical, the
essentially primitive associations of the country dance in 'East Coker'
lead it to be classed with the 'folly' of Frazerian savages; its fertility
rhythm, 'Nourishing the corn' like that of the inescapably cyclic
fertility ceremonies of *Sweeney Agonistes* and *The Waste Land*,
points towards 'Dung and death'. Admired and rejected, the primitive
rite is also linked with the city; the isolated last line of the poem's
second section summons up the field and its dancing, leading to a
juxtaposition with the urban world which follows.

---

[101] 'A Commentary', *Criterion*, Oct. 1926, p. 628; 'In Praise' (C478), p. 156.
[102] See Gardner (1978), p. 99.
[103] Gardner (1978), p. 99; see also Vickery, pp. 276-7.

The dancers are all gone under the hill.

III

O dark dark dark. They all go into the dark,
The vacant interstellar spaces, the vacant into the vacant,
The captains, merchant bankers, eminent men of letters.
. . .
Distinguished civil servants, chairman of many committees, . . .

As with the living dead of *The Waste Land*'s London and the inane
savages of its 'mudcracked houses', so in 'East Coker' the primitive
rituals of the countryside and the (essentially commercial) life of the
city and City both point only towards death. But now Eliot clearly
finds aspects of London life and country life congenial. Having
sympathetically implicated elements of his East Coker ancestry and
private 'mirth' in the opening section, here the poet, himself content
to sit on 'many committees' and regarded as one of the 'eminent men
of letters', is implicated in the world he attacks. The circles of the
dance are as invalid as the circles of the Underground. Their form
or pattern, dead in itself, may hint at something higher, but unless
that higher truth is realized, they remain meaningless and deathly.
Both East Coker and London were in a sense 'home' for Eliot,
but, as he reminds us 'Home is where one starts from'. One of the
uses of understanding how this transform of the city and savage
theme continues in Eliot's work is that in seeing how this primitive
countryside (like the animistic rose-garden) connects with London's
metropolitan world, it becomes apparent that there is a political
aspect to the *Four Quartets*. The politics are closely related to Eliot's
prose writing of the period. He recalls how, seeing England from
his own urban American background he has always been aware that
the natural habitat of the Englishman was the little rural community
and that industrial and large towns were accepted with reluctance.
Eliot sees London as unusual among big cities in having grown up
gradually out of a grouping of villages and concludes that 'there is
something about England which remains stubbornly attached to the
parochial'.[104] As the very titles of the *Quartets* remind us, Eliot is
stressing the importance of this parochial tradition. Significantly, it
is here that he locates 'mirth'. Yet the overt ruralism of the prose
is more strongly qualified in the verse, where country, primitive, with

---

[104] 'The English Tradition', *Christendom*, Dec. 1940, pp. 226-7.

its 'daemonic, chthonic / Powers' is seen as ultimately no better than city unless redeemed by the Christian vision. And in poems which make use of primitive ceremony, several foreign literatures, and much Buddhist thought, Eliot's parish is no Little England. The *Quartets* are all the more impressive for having this political dimension worked into them, yet still subordinated beneath the religious scheme.

Their homecoming, going to earth and returning to sources, reaches, in a sense, its furthest point in 'The Dry Salvages', where Eliot returns to his own St Louis and Massachusetts childhood, to 'The life of significant soil', and to the savage meanings associated with such beginnings. The poem stresses throughout the elemental qualities of the landscape and seascape which it describes, leading Eliot to a particularly bare group of rocks which he had known from the sailing days of his childhood. But the three rocks also attract him because of their name (see p. 34 above).

It is with this savage, primitive element that the poem begins:

> I do not know much about gods; but I think that the river
> Is a strong brown god

Eliot's familiar theme that the savage exists even below the urban world on a forgotten, yet still potent and often dangerous level is here reiterated:

> The problem once solved, the brown god is almost forgotten
> By the dwellers in cities—ever, however, implacable,
> Keeping his seasons and rages, destroyer, reminder
> Of what men choose to forget.

'Within us' the Mississippi in St Louis, like the voices in the rose garden in 'Burnt Norton', or the dancers in the field of 'East Coker', or the village concealed in the shape of modern London, or the savage elements in modern man, maintains a secret presence, 'waiting, watching and waiting'. It has been present from the beginning as a primitive rhythm, that element associated with the 'auditory imagination' of poetry, and so with the savage in the jungle and his gods, 'His rhythm was present in the nursery bedroom . . . '. Primitive and child are again linked, but not as the second childhood of 'The Hollow Men'. The primitivism of the 'strong brown god' is healthier: a childhood promising growth. Prefacing *Huckleberry Finn* (first read 'a few years ago') in 1950, Eliot recalled his own boyhood. His words suggest that Twain's book brought back Eliot's childhood not merely by functioning as an *aide-mémoire*. It enabled Eliot to look not to the detested Unitarianism of his family, but towards a deeper, wider, yet also more personal 'rhythm' present, though scarcely recognized at the time. Eliot, thanks to Twain, is able to see his childhood as containing, in Durkheim's phrase, *The Elementary Forms of the Religious Life.*

. . . the River makes the book a great book. As with Conrad, we are continually reminded of the power and terror of Nature, and the isolation and feebleness of Man. Conrad remains always the European observer of the tropics, the white man's eye contemplating the Congo and its black gods. But Mark Twain is a native, and the River God is his God. It is as a native that he accepts the River God, and it is the subjection of Man that gives to Man his dignity. For without some kind of God, Man is not even very interesting.[105]

So accustomed to using Conrad's eyes, Eliot is able to see through Twain's the native emotion of his own nativity and childhood in the nursery bedroom, seeing that childhood in terms of a fundamental primitivism from which his religious consciousness has grown. Here again, anthropology and romanticism have coalesced.

The experience represented by river and sea is the experience of tradition's endurance, and a tradition, even in its full development, has no truck with 'superficial notions of evolution [as] a means of disowning the past'. Rather, the tradition includes and is aware of its primitive origins: this reminds the tradition of its vulnerability and need to look for higher guidance, while keeping it aware of crude root emotions which cannot be avoided. The fear of savage 'folly' links with the fear of God.

I have said before
That the past experience revived in the meaning
Is not the experience of one life only
But of many generations—not forgetting
Something that is probably quite ineffable:
The backward look behind the assurance
Of recorded history, the backward half-look
Over the shoulder, towards the primitive terror.

Eliot's intense sympathy with the fishermen of the Massachusetts coast transforms them into figures of endurance, secular saints,

---

[105] 'Introduction' to *Huckleberry Finn* (B59), pp. vii and xv.

whose course is one of earthly failure which the toughest faith transforms into gain. This daily confronting of the 'primitive terror' as represented by the sea is held up for our admiration. There may be an eastern quality to such reverence for long, ancestral tradition; this section is followed by the episode of Krishna and Arjuna which is related to the business of the fishermen by the concluding 'Not fare well, / But fare forward, voyagers.' All this contrasts sharply with the flimsy world of divination, of Madame Sosostris, which lands us unsurprisingly in the heart of London as we hear how all this 'fiddle' will always be found 'When there is distress of nations and perplexity / Whether on the shores of Asia, or in the Edgware Road'. What raises Eliot above the entrapping circles of city and savage is what for him has continually to raise words towards the Word, something made available in partial revelations. 'The hint half guessed, the gift half understood, is Incarnation.' Here city and savage are united and transcended. We are left with a verdict which sees the savage as fatally limited because cut off from God's Word which passes understanding; similarly, *The Waste Land*'s fertility cults were cut off from the peace of 'Shantih shantih shantih'. Eliot's leavetaking of the city comes in 'Little Gidding'. In 'The Dry Salvages' he salutes and passes the 'daemonic, chthonic / Powers' of the savage.

Charles I, Christ, and Frazer's monarchs coalesce in the 'broken king' of 'Little Gidding' whose 'May, with voluptuary sweetness', recalling Gerontion's 'depraved May', perhaps follows an urge to juxtapose spring's arrival with the breaking of the king.

What matters more is the poem's preoccupation with the city and its destruction. If a savage element persists, it is the modern savagery of the Blitz which, physically, as a firewatcher, Eliot attempted to assuage. The sight triggered deeply his apocalyptic imagination. What had happened in verse at the conclusion of *The Waste Land* was actually happening: London was being destroyed.

At first in 'Little Gidding', the bombing and the burning seem to be happening in the countryside, part of that cycle of death and rebirth opening 'East Coker', explicitly recalled.

> Dust inbreathed was a house —
> The wall, the wainscot and the mouse.
> The death of hope and despair,
>   This is the death of air.

The cycle of death leads us on towards the urban landscape that follows. The 'Dust in the air suspended' seems to present us with a barren desert of death. Eliot had again used the traditionally romantic similar alternatives 'a desert or a city'. Yet, as one of his prose pieces of the period reveals, the desert remained for him not only a place of death, but also a place of Christian triumph. He gave high praise to the biography of a French priest killed in North Africa. Charles de Foucauld travelled disguised, the first explorer 'in unexplored and unsubdued territory'. He gave aid to the tribesmen and aimed primarily not to convert by teaching, 'but to *live* the Christian life, alone among the natives'. Finally, Foucauld was killed by a marauding band which was unaware of his name and reputation. 'There is no higher glory of a Christian empire than that which was here brought into being by a death in a desert.'[106] This piece relates to the 'explorers' of 'East Coker', and to the poem on 'the Indians who Died in Africa'. It is the revision of the theme of cannibal and missionaries in *Sweeney Agonistes* which makes possible the fate of Celia in *The Cocktail Party*. It also relates to Eliot's sense of the poet's mission in 'Little Gidding' when the streets of London become simply a tribal homeland.

In the city, 'In the uncertain hour before the morning', Eliot meets his 'familiar compound ghost'. The use of Dante is encouraged by *The Waste Land* and by *The City of Dreadful Night*. The time for Eliot is just after the passing of 'the dark dove with the flickering tongue', a German plane which has left behind a city every bit as much of a waste land as the earlier poem's London, though now the 'crowd' has gone, leaving only the speaker who, 'before the urban dawn wind', communicates with the ghost.

> Since our concern was speech, and speech impelled us
>   To purify the dialect of the tribe . . .

Typically, the second line is not Eliot's own but is selected deliberately. The English language, to which Eliot had dedicated (vainly, he worried) so much of his life is here reduced merely to 'the dialect' of a 'tribe'. As dialect, it is all the more vulnerable to disintegration and collapse. Similarly as a 'tribe', rather than a nation, the English (and humanity) are revealed as more vulnerable. The effect also forges a link with those primitive origins of poetry, which 'begins, I dare say, with a savage beating a drum in a jungle'. This conversation

---

[106] 'Towards a Christian Britain', *Listener*, 10 Apr. 1941, p. 525.

at the threatened end of a city and a civilization links its poet with
the primitive, savage beginnings. The speech is filled with self-
reproach, yet it ends with what was promised at its start: dawn. Day
rises over the broken city.

> The day was breaking. In the disfigured street
> He left me, with a kind of valediction,
> And faded on the blowing of the horn.

All is clear. The final, Jericho-like sound may be frightening, but
is at least an escape from

> The sound of horns and motors, which shall bring
> Sweeney to Mrs. Porter in the spring.

The saving of the poet demands 'that refining fire / Where you must
move in measure, like a dancer'. Similarly, the saving of the city seems
to demand its destruction. Eliot's apocalyptic urban imagination had
destroyed London before. Now it must be done again. There is both
an exultant and a suffering note in the return of the destructive dove
which is also the Christian bird of Incarnation whose message for
both individual and society, parish and city, is driven home in the
famous anthem passage when 'The dove descending breaks the air.'
The God behind this love is a savage God as well as a redeeming,
loving one. After this passage, there is no more city, no more savage
in the *Quartets*. Only the return through poetry and childhood
images to the final home which lies in the blank page following the
moment when 'the fire and the rose are one'.

In Eliot's next prose book and in his next play, the combined themes
of city and savage were important. *Notes towards the Definition of
Culture* supplements Arnold's definition not by abandoning it, but
by setting it beside a wider, basically anthropological definition of
culture. The essentially anthropological basis of Eliot's most crucial
definition of culture is clear in the published book, as Bantock
indicates, but the earlier (1943) version of the first section reveals
the intellectual movement even more clearly.[107]

Eliot sets aside Arnold's notion of culture, stating that his own main
interest is 'with the culture that a whole society may develop and
transmit'. Yet the first definition of culture which he now supplies,

---

[107] G. H. Bantock, *T. S. Eliot and Education* (London: Faber & Faber, 1970),
p. 64.

'that of a refinement of living, including appreciation of philosophy and the arts, among the upper levels of society', is in many ways close to Arnold's ideal. In Eliot's own life such an idea seems to have been associated particularly with artistic movements of the big cities. Against this first idea of culture he sets a second definition at once wider and 'more fundamental'.

. . . it is what we imply when we speak of *Primitive Culture*. It is the whole complex of behaviour, thought and feeling, expressing itself in custom, in art, in political and social organisation, in religious structure and religious thought, which we can perceive most clearly as a whole in the less advanced societies, but which is equally present as the peculiar character of the most highly developed people or nation.[108]

There is not detailed remembering of the book, but even the full title, *Primitive Culture: Researches into the Development of Mythology, Philosophy, Religion, Language, Art, and Custom*, shows that Tylor's conception of culture as all embracing is one of the keys to Eliot's work. From the start, his two definitions of culture, that of a class and that of the whole people, 'have to be kept distinct but always in relation'. Particularly in the music-hall connections of his plays (which later became their West End connections) we can see that he was attempting to do this in his own creative work at the time. *Notes towards the Definition of Culture* takes the 'savage' as a base. Though Eliot refuses to treat the uneducated merely as 'savages', his essential ideas about the indispensable contribution of religion to culture are presented as being derived from the study of a less complex model. The original article shows clearly his move from primitive to sophisticated, as he demonstrates the way in which amongst primitive communities various parts of culture are inextricably interlaced. Eliot instances the activities of Dyak head-hunters the production of whose carvings is at once a task utilitarian and sacred. As civilization develops, though, occupational specializations proliferate, and eventually there appears a capacity for abstracting religion, science, politics, and art from one another. Eliot moves from such a primitive organization to discussing Greek drama, following the movement of Harrison, Cornford, and the other anthropologically influenced classical scholars whom he had read.

---

[108] 'Notes towards a Definition of Culture, I', *New English Weekly*, 21 Jan. 1943, p. 117.

He sees that conflict between various elements in highly developed society can be creative, but that eventually deterioration will follow if connections are not preserved between the various areas of life which, in the primitive model, are integrated. Without naming names, he goes on to outline the situations which had so interested him in the cases of the Melanesians and the Tari Furora, as he points out that to tamper with the pattern of primitive culture at one point is to endanger the whole structure.[109] He relates this to the state of culture of his own age, facing dangers of over-specialization, which impoverishes both the religious and artistic sensibilities by separating each from the other, so that only 'the vestige of *manners* may be left for those who, having their sensibility uninformed either by religion or by art, . . . have nothing left but an inherited behaviour which ceases to have meaning'.[110] Eliot had written of and against such a *milieu* when he combined the elements of the savage and the city, from the beating tom-tom of 'Portrait of a Lady' through the rituals of *The Waste Land* and *Sweeney Agonistes* to the cocktail chatter and jungle martyrdom of the play, *The Cocktail Party*, yet to come. Revising the original articles for *Notes towards the Definition of Culture*, he complicated his argument's texture by involving more material relevant to his personal history and to the history of his work, such as that mention of *Heart of Darkness* which looks back to *The Waste Land*.[111]

The first article in the series which became the book demonstrated that Eliot's definition of culture begins with the savage. The second showed that at the other pole is the city. Eliot discussed the shrunken sense in which 'culture' was applied to the arts, culture centring on Paris, London, and more recently New York. While the city is kept within bounds he sees that this can be healthy, though in both culture and industry a balance between town and country is necessary. As in *Four Quartets*, it is the country which has inherited the virtues of primitive community, the savage's integrated civilization being reflected in what Eliot, following fashion of the day, praises as organicism. 'Organicism' suggests close-knit near-to-the-soil values of rural communities whose established tradition and integrated value-system are threatened by the sophisticated metropolitan. Eliot's thinking here sometimes relies on romantic clichés insufficiently analysed in Russell Square, yet it could also produce provocative

---

[109] Ibid.   [110] Ibid., pp. 117–18.   [111] *Notes*, p. 41.

insights. The urge to preserve these rural communities led him from examination of dying primitive societies to sympathy with developing regionalism as a political force to preserve cultural values under threat. Such regional movements, whether in Scotland or America, tended to be associated with agrarian, traditional integrated societies beyond the 'wholly urban-minded' capital.[112]

Eliot made clear in *Notes* that his true concern was 'a problem of the first importance . . . that of the *transmission of culture*'.[113] The book's scope extends far beyond consideration of Karl Mannheim's stress on a need for 'elites'. Eliot's third article again stressed that linkage of primitive and sophisticated, first linked explicitly in the review of Wyndham Lewis's 'cave-man' novel in 1917. Eliot did not want 'a romanticized folk-culture'. Rather, 'the censure of this essay must be on a somewhat different ground—that of my wishing to have all at once what can only be had in succession in an historical process, a co-existence of the primitive and the most highly sophisticated'. Here it seems Eliot's artistic practice is clearly carrying over into his social thought. His reply to the expected criticism is that ideals must be striven for particularly now since 'in a position in which old peoples and old cultures are likely to be supplanted by new—we are in a common danger to all races over the whole circumference of the globe'.[114]

This global stance and anthropological imagination continue to guide him in his consideration of the relation between religion and culture, and his fascination with 'more primitive and self-contained peoples, where the culture and the religion are co-terminous'.

Wishing for 'much the same unity on a higher plane', he presents himself as a kind of agriculturalist, concerned with 'our relation to the spiritual soil'. Against those iconoclasts who would rid life of all images in the name of religion, he sees repudiation of art as acceptable for the extreme ascetic, but, if universally applied, as starving the spiritual soil of nourishment. Yet he insists that if the arts are a necessary element in culture and if culture is essential to develop a people's highest spiritual capacities, then it must not be

---

[112] 'Notes towards a Definition of Culture, II', *New English Weekly*, 28 Jan. 1943, p. 130.
[113] *Notes*, p. 40.
[114] 'Notes towards a Definition of Culture, III', *New English Weekly*, 4 Feb. 1943, pp. 136–7.

forgotten that without religion no culture can exist.[115] This is the main message of these 'Notes'. Whether the reader disagrees with that message or not, Eliot cannot be accused of suddenly presenting it or of bringing it to the fore simply because of his Christian faith. Certainly it accords with and draws on that faith, but also and explicitly it comes from Eliot's long concern with the meeting extremes of savage and city.

I have concentrated on the articles, 'Notes towards a Definition of Culture', rather than the less tentatively titled *Notes towards the Definition of Culture* because the book, while refining and adding to the argument, occludes that argument's clarity through continual divagation and a tendency to over-engage in discussion with Eliot's own contemporaries, many now forgotten. A similar tendency vitiates *The Idea of a Christian Society*, with its many and lengthy endnotes. The defence of the format of these books is that they were written as part of a struggle. Now that details of that struggle in the later 1930s and 1940s have faded, the books have been weakened, though their central argument can still engage and provoke.

In various articles of the forties Eliot refined his material, whether in relating poetry to religious ritual in 'The Social Function of Poetry' or returning to *Primitive Culture* and John Layard's studies 'in the "stone age" New Hebrides' in 'Cultural Forces in the Human Order' which reworks many of the points in the earlier 'Notes' series.[116] Simultaneously, he concerned himself with the idea that cities were necessary, not evil, but that 'without the life of the soil from which to draw its strength, the urban culture must lose its source of strength and rejuvenescence'.[117] He did not always side with the city, though he did concern himself with urban and industrial problems. Writing on 'Full Employment and the Responsibility of Christians' (1945) in a response to a correspondent signed 'Civis', Eliot signed himself 'Metoikos'—resident alien.[118] He continued to set out ideas on

[115] 'Notes towards a Definition of Culture, IV', *New English Weekly*, 11 Feb. 1943, pp. 145–6.
[116] 'The Social Function of Poetry', *Norseman*, Nov. 1943, p. 450; 'Cultural Forces in the Human Order', in Maurice B. Reckitt, ed., *Prospect for Christendom* (London: Faber & Faber, 1945), pp. 57 and 59.
[117] 'The Responsibility of the Man of Letters in the Cultural Restoration of Europe', *Norseman*, July/Aug. 1944, p. 244.
[118] 'Full Employment and the Responsibility of Christians', *Christian News-Letter*, 21 Mar. 1945, p. 12.

cultural issues in a piece on 'The War and the Blitz' also incorporated into the *Notes*. Again he pointed out 'more marked differentiation of function' in higher types of primitive society.[119] Writing about UNESCO in 1947, he repeated that he 'had always understood culture to comprehend education and science, as well as other interests'.[120]

Prefacing *Notes towards the Definition of Culture*, Eliot acknowledged particular debts to V. A. Demant, Christopher Dawson, and Karl Mannheim.[121] Mannheim, whose work Eliot read after 1937 and whom he frequently met, has been discussed by Kojecky.[122] For the themes of savage and city it is Demant and Dawson who are important, though neither writer is gripping. Dawson, the more significant figure, had featured in the *Criterion* since the late 1920s, Demant since the early 30s. Dawson's first book, *The Age of the Gods* (1928), subtitled *A Study in the Origins of Culture in Prehistoric Europe and the Ancient East*, began by noting that 'During the last thirty years the great development of archaeological and anthropological studies has prepared the way for a new conception of history.' Dawson was interested in history 'not as an inorganic mass of isolated events, but as the manifestation of the growth and mutual interaction of living cultural wholes'. For Dawson, organic connections were good. He emphasized what is summed up in the title of Simone Weil's book which Eliot would later preface, *The Need for Roots*.

. . . the earliest agriculture must have grown up round the shrines of the Mother Goddess, which thus became social and economic centres, as well as holy places, and were the germs of the future cities.

Dawson emphasized the importance of combining agricultural and urban for building a strong society. He also discussed 'savages', referring to figures such as Frazer, Tyler [*sic*], Rivers, Elliot Smith, Marett, and Durkheim. Often Dawson's ideas are remarkably similar to those of Eliot.

A culture can only be understood from within. It is a spiritual community which owes its unity to common beliefs and a common attitude to life, far more than to any uniformity of physical type.

---

[119] 'The Class and the Elite', *New English Review*, Oct. 1945, p. 499.
[120] 'UNESCO and the Philosopher', *Times*, 20 Sept. 1947, p. 5.
[121] *Notes*, p. 9.
[122] Roger Kojecky, *T. S. Eliot's Social Criticism* (New York: Farrar, Straus and Giroux, 1971), esp. ch. X.

Hence the study of primitive culture is intimately bound up with that of primitive religion.[123]

Dawson's 1932 book, *The Making of Europe*, drew his work into a more specifically European context, emphasizing classical tradition, the importance of tribal as well as urban culture, and the significance of asceticism, particularly as shown by the desert saints. Carol Smith has pointed to the importance of this book for the primitive ritual elements in Eliot's drama.[124] It contributed also to his stress on 'The Unity of European Culture', though ever since 'Tradition and the Individual Talent' Eliot had been preoccupied with 'the mind of Europe'. Dawson underlined the importance of the family, a concept which became more and more explicit not only in Eliot's later social writings but also in his plays. Like Eliot, Dawson emphasized the 'two Englands' created in the nineteenth century—'the England of the fields and the England of the factories'—and wished to build on a common 'English tradition' which with some sort of religious sanction would take people *Beyond Politics*.[125]

In many ways on the outbreak of war this was, of course, a naïve position. It was also profoundly necessary to understand roots and strengths of the culture so that what was best might be salvaged from carnage, and worries like Eliot's—that the war might be fought only to preserve financial interests—could be assuaged. The positions of Dawson and Eliot moved closer and closer together, influencing one another. The two men conversed. Dawson's *Religion and Culture* stressing Tylor's importance and relating theories of primitive religion to Christianity was selected by Eliot as one of his 'Books of the Year' in 1950.[126]

V. A. Demant, a parish priest as well as a Fellow of the Royal Anthropological Society, also stressed 'The Importance of Christopher Dawson'.[127] Demant too combined interest in savage and city. He cites Malinowski's *Sex and Repression in Primitive Society*, criticizing

---

[123] Christopher Dawson, *The Age of the Gods* (London: John Murray, 1928), pp. v, 11, 382, 22–27, 22.

[124] Carol Smith, p. 83.

[125] Christopher Dawson, *Beyond Politics* (London: Sheed and Ward, 1939), pp. 45 and 54.

[126] Kojecky, pp. 238–9; 'Books of the Year Chosen by Eminent Contemporaries', *Sunday Times*, 24 Dec. 1950, p. 3.

[127] V. A. Demant, 'The Importance of Christopher Dawson', in *Theology of Society* (London: Faber & Faber, 1947), pp. 185–99.

the way in which 'the fashionable psycho-analytic school of Sigmund Freud deduces the processes of the human psyche from the particular problems of the modern urbanized European'.[128] Demant's 'Christian Sociology' was important not just in itself; it threw up names from Eliot's past, such as those of Maine and Durkheim, which maintained the importance of such figures within an explicitly Christian context. The third section of Demant's 1936 *Christian Polity* (Faber & Faber) is given over to anthropological essays, such as 'Idols: Their Place and Influence in Religious History' where again we meet the names of Elliot Smith, Tylor, Ellis, Frazer, McDougall, Robertson Smith, and others.[129] Eliot now tended to be interested in anthropology mainly when it contributed to a Christian perspective, so that while he would maintain that 'the actual religion of no European people has ever been purely Christian, or purely anything else' since 'There are always bits and traces of more primitive faiths, more or less absorbed', he became more and more interested in relating such faiths to Christian problems. So he was attracted to Evans-Pritchard's contention that 'the sociologist should also be a moral philosopher and that, as such, he should have a set of definite beliefs and values in terms of which he evaluates the facts he studies as a sociologist'.[130] Evans-Pritchard's 'Social Anthropology' devotes much attention to the Victorian anthropologists, Tylor and Frazer, then passes on to their critics, 'Durkheim and the *Année Sociologique* group of writers', thus recapitulating Eliot's own interests; some 'features may be found to exist in all human societies, primitive and civilized alike'; the paper relates anthropology to religious belief, 'with special reference to Catholic apologetics'.[131] This concurrence of anthropology and Christian belief was, similarly, what attracted Eliot to Demant, who was editor of the 1944 lectures, *Our Culture: Its Christian Roots and Present Crisis*. Demant's own chapter would have been all the more attractive to Eliot because of its wide-ranging view which combined the primitive and the sophisticated.

As we might speak of the culture of the Hebrides, or as W. H. Rivers wrote of the culture of Melanesia, so we speak of the culture of the European West

[128] V. A. Demant, *God, Man and Society* (London: Student Christian Movement Press, 1933), p. 165.
[129] V. A. Demant, *Christian Polity* (London: Faber & Faber, 1936), pp. 183–206.
[130] *Notes*, pp. 31–2 and 69 n.
[131] E. E. Evans-Pritchard, 'Social Anthropology', *Blackfriars*, Nov. 1946, pp. 409–14.

to describe a set of outlooks, aims and ways of life, which has a history in the past of Europe and has spread to other parts of the world.[132]

Again, there seems to have been cross-fertilization between Demant's work and that of Eliot.

The Cocktail Party, whose first draft of three scenes was completed by June 1948, more than any other of Eliot's later plays unites the themes of savage and city.[133] This is hardly surprising, since its gestation period comprised the years in which Eliot was working on the *Notes*. The sharp combination of savage and city along with ritual and music-hall elements do much to make this the most powerful of Eliot's comedies. '*The scene is laid in London*,' in the tinkling, surfacy cocktail world of sophisticated chiming dialogue more realistic than that of *Sweeney Agonistes*, but still reminiscent of the earlier play in the chinking rhythms of its conversation.[134] *The Cocktail Party* presents again a world whose moral base has been eroded, though surface conventions remain. Into it comes the shamanic figure of 'An Unidentified Guest' who is in control of potent and primitive forces, which the other characters do not seem to understand. Reilly's knowledge seems to surpass the barriers of the natural. Associated with the Devil by Celia (I. ii), he is a sort of devil doctor.

> UNIDENTIFIED GUEST. . . . to approach the stranger
> Is to invite the unexpected, release a new force,
> Or let the genie out of the bottle.
> It is to start a train of events
> Beyond your control.[135]

Straddling naturalistic setting and ritual drama, Reilly is the only character who sings. His song seems like a recollection of one of Eliot's music-hall favourites and heightens the unusual, ritual element of his drinking. The play's strength is that it draws on Eliot's earlier work and makes that earlier work transferable to the West End, a triumph in itself; its weakness, like that of most of the plays, is that it offers us little we cannot find more concisely and intensely expressed in the poetry. Edward's speech, for instance, presented

---

[132] V. A. Demant, ed., *Our Culture* (London: Society for Promoting Christian Knowledge, 1947), p. 1.
[133] See Eliot's letter of 1 June 1948 in Browne, p. 172.
[134] *CPP*, p. 352.   [135] *CPP*, p. 361.

'Contre Sartre', and stating that 'Hell is oneself', recalls 'Thinking of the key, each confirms a prison', and Bradley on 'the whole world for each' as 'peculiar and private to that soul'.[136] Other earlier themes resurface. As in 'Marie Lloyd' the growth of the cinema was juxtaposed against the primitive Melanesian life, so in the play the interest of Peter Quilpe and Celia Coplestone in Hollywood contrasts with the reality of the primitive world for which, eventually, Celia abandons the Hollywood dream. Unreality and reality continue to cross and recross throughout the play, as do the savage and the sophisticated. London chatter, beginning with tigers in India, involves themes such as Montenegro peasants, primitive Guardians, libation rituals, and unrest among natives as well as more expected Hampstead material. The matter of the drama continues to be eclectic as Eliot pokes beneath the glossy city surface to hidden depths of feeling, and as so often looks to Buddhism as well as Christianity. Those last words of the Buddha, which are also contrived as Reilly's last words to Celia — 'Work out your salvation with diligence' — look back to Eliot's 1914/15 essay discussing the supremely attainable good.[137] But the final juxtaposition, and the final breakthrough to illumination and an absolute religious certainty, comes about through a direct confrontation between the savage and the city, which proved shocking to the original audiences and retains some of its power to shock today. Celia's death is the equivalent of the death in the desert undergone by Foucauld. It shows Christianity beside the most primitive of humanity and slashes a raw strength into the London lounge. The confrontation is introduced by being absurdly dismissed, in a West End version of *Sweeney Agonistes*, with a small side glance at Lévy-Bruhl's prelogical mentality thrown in as 'The native is not, I fear, very logical.'[138] From slick dialogue about eating Christians, the play moves back to the world of the cinema. Peter has just come from California to make 'a film of English life' which will involve the reconstruction of an English residence in Hollywood. He wished Celia to star in his film, but the unreality of this world is punctured by the stark reality of the ritual slaughter. Celia perished trying to help some natives dying of pestilence. Only traces of her body have been found.

---

[136] Eliot quoted in Browne, p. 233; *CPP*, pp. 397, 74, and 80.
[137] *CPP*, p. 420; 'Ethics', unpublished paper (Houghton), p. 21.
[138] *CPP*, p. 429.

ALEX.   It was difficult to tell.
But from what we know of local practices
It would seem that she must have been crucified
Very near an ant-hill.[139]

Initially the description of Celia's death had been more savagely
horrifying, describing the body's 'decomposition' and the way in
which the tribesmen 'smear the victims / With a juice that is attractive
to the ants'.[140] Because of shock caused among audiences by this
last line, Eliot was persuaded to alter the passage, so that it is clear
that shock is only a part of a deeper message which shows how
contact with the primitive can renew values of faith, self-sacrifice,
and idealism which seem to have become totally atrophied in the
world of London cocktail party and Hollywood film. The play's
message then is fundamentally religious. The London socialites come
to seem no more important than the savages.

REILLY.   Who knows, Mrs. Chamberlayne,
The differences that made to the natives who were dying
Or the state of mind in which they died?[141]

Christianity is largely defined against the savage world, though
the savage world is seen as an essential base and background. Eliot
has been careful to avoid any sort of sentimental religiosity, and
this strengthens greatly the play's climax. Yet it is interesting to
read another speech, reluctantly dropped very late in the play's
completion, which both showed Christianity in some ways working
with the savage world, and looked back with a different point of
view to the earlier worship of a human god in the unpublished
'Exequy' poem in the *Waste Land* manuscripts.

LAVINIA.   . . . Yet I thought your expression was one
of . . . satisfaction!
Interest, yes, but not in the details.
ALEX.   There's one detail which *is* rather interesting
And rather touching, too. We found that the natives,
After we'd reoccupied the village,
Had erected a sort of shrine for Celia
Where they brought offerings of fruit and flowers,
Fowls, and even sucking pigs.

[139] CPP, p. 434.   [140] Earlier version cited by Browne, p. 226.
[141] CPP, p. 436.

> They seemed to think that by propitiating Celia
> They might insure themselves against further misfortune.
> We left *that* problem for the Bishop to wrestle with.[142]

Martin Browne's question, 'Do we hear echoes of "the glittering, jewelled shrine" of Thomas Becket?' may be a little too direct. But certainly here, for the clearest moment in Eliot's later plays the savage and the Christian are joined in reproach of the oversophisticated city; though it is essential, too, to remember that that city also contains Sir Henry Harcourt-Reilly whom Eliot once described as possibly 'a god *in* the machine'.[143]

It is possible to see faint remnants of the city/savage confrontation in the later plays, and Carol Smith continues to track the ritual elements in these, but the feeling in them seems to me increasingly one of whimsy and, eventually, gentle pastoral. The confidential world of the city clerk persists in Eliot's poorest play which looks again to Victorian melodrama. Generally, though we may have the background threat of Gomez's South American violence or past humiliation (Maisie Mountjoy brings back the music hall again), the late plays are too conventional to take us again immediately from London to Kinkanja. When they return to sources they do so in an increasingly normal, quietly acceptable manner that brings a more domestic way of ending.

> I feel utterly secure
> In you; I am a part of you. Now take me to my father.[144]

This is in keeping with the conclusion of Eliot's own life.

These late plays show a continuing interest in moments when 'life is elevated to the dignity of dance or liturgy, with a gaiety', which Eliot (perhaps with a glance at Yeats's 'Lapis Lazuli') now sees as 'in all great poetry, and the greater seriousness behind the gaiety. It is in fact the privilege of dramatic poetry to be able to show us several planes of reality at once.'[145] In the distance, still, is *Sweeney Agonistes*.

The late prose shows occasional glimpses of themes discussed earlier. But these later pieces are of most use for the light they shed

---

[142] As n. 140 above, p. 227.
[143] T. S. Eliot and Iain Hamilton, 'Comments on T. S. Eliot's New Play, *The Cocktail Party*', *World Review*, Nov. 1949, p. 21.
[144] *CPP*, p. 583.
[145] 'The Aims of Poetic Drama', *Adam International Review*, Nov. 1949, p. 16.

on the earlier development. Sometimes perhaps unlikely sources of influence are revealed, as when Eliot quotes with interest Gordon Craig's statement that 'The father of the dramatist was the dancer,' or continues to admire the combination of 'art, ritual and music' now seen as 'always the handmaid of the beauty of holiness'.[146] Though he now said that he was 'no longer very much interested in my own theories about poetic drama, especially those put forward before 1934', the old interests which had fascinated him from his first dramatic *Fragments* continued to grip him, leading to the fact that each of his dramas had as its 'sort of springboard' a 'Greek myth'.[147] In 1955 he wrote to Philip Mairet, to whom he had dedicated *Notes towards the Definition of Culture*, stressing that he saw ritual as an essential element in life.[148] Eliot had not altogether left behind his earlier anthropological interests. 1956 saw him complaining that the author of an article on 'Christian Social Thought' had used the word 'sociology' so that 'unless my memory is greatly at fault, Durkheim and Lévy-Bruhl' would be excluded from its province.[149] In these years he remembered, too, his early concern with the city, recording his debts to Thomson and Davidson.[150] In 1958, Eliot was still pointing out the horror of a civilization whose materialism is 'simply boredom'.

A people without religion will in the end find that it has nothing to live for. I did touch on this problem a good many years ago in an essay I wrote on the death of a great music-hall artist, Marie Lloyd.

He was musing that the synthetic religions of Stalin and Hitler should neither of them 'properly be called pagan, but if you do call them pagan then we must say that they're inferior as religions to genuine primitive pagan religion'.[151] Proof that the themes of savage and city stayed in Eliot's mind to the end is furnished by one of the last pieces which he published. In his 'Preface' to Edwin Muir's *Selected Poems*, he gives high praise to a poet whose work is very different from his own. Finally, he is drawn to the essence of Muir's genius

---

[146] 'Gordon Craig's Socratic Dialogues', *Drama*, Spring 1955, p. 18; 'From Gloucester Road', *Church Times*, 9 Mar. 1956, p. 12.
[147] 'The Art of Poetry' (C631), p. 62; p. 61.
[148] Letter to Philip Mairet; 23 Mar. 1955 (Austin, Texas).
[149] 'Christian Social Thought', *TLS*, 20 Apr. 1956, p. 237.
[150] 'Preface' to *Davidson* (B83) p. xi.
[151] Leslie Paul, 'A Conversation with T. S. Eliot', *Kenyon Review*, Winter 1964/65, pp. 14 and 12.

which he sees as that of 'the sensibility of the remote islander', and the words that follow give one final transformation of the savage and city motif, when he describes Muir as 'the boy from a simple primitive offshore community who then was plunged into the sordid horror of industrialism in Glasgow, who struggled to understand the modern world of the metropolis in London . . . '[152] Like Muir, Eliot had won through to a vision of final acceptance, which allowed him to look back even to Sweeney and to call him, at Columbia in 1958, 'friend'.[153] This informal remark shows an inescapable attachment to a character epitomizing one of the most fascinating, longest-lasting, and most potent aspects of Eliot's work: its binding together of the savage and the city.

[152] 'Preface' to Edwin Muir, *Selected Poems* (London: Faber & Faber, 1965), p. 10.
[153] TS of notes later published as 'T. S. Eliot talks about his poetry', *Columbia University Forum*, Fall 1958, p. 13 (Butler). In the published version, Sweeney is simply a 'man'.

SELECTIVE BIBLIOGRAPHY

1. UNPUBLISHED WORKS

Many hundreds of unpublished items were examined. The following major collections were particularly useful: Beinecke Rare Book and Manuscript Library, Yale University; Berg Collection, Astor, Lennox, and Tilden Foundations, New York Public Library; Bodleian Library, Oxford; British Library, London; Brotherton Collection, University of Leeds; Butler Library, Columbia University; Houghton Library, Harvard; Huntington Library, California; King's College Library, Cambridge (Hayward Bequest); Library of the Humanities, Austin, Texas; Pusey Library, Harvard. Though not specifically cited, material from collections at the following institutions was also useful: Balliol College Library, Oxford; General Manuscript Collection, New York Public Library; Maclehose Collection, Mitchell Library, Glasgow; Matthiessen Room, Eliot House, Harvard; Merton College Library, Oxford; National Library of Scotland, Edinburgh; Pepys Library, Magdalene College, Cambridge; Princeton University Library.

2. INTERVIEWS

Particularly useful are the interviews contained in 'The Mysterious Mr Eliot' (BBC TV documentary, January 1971). In addition, I am grateful to the following people who generously shared their memories of Eliot: Professor William Alfred, Professor Richard Ellmann, Professor John Finlay, Dr Donald Gallup, the late Dame Helen Gardner, Professor Harry Levin.

3. WORKS BY ELIOT

Almost all of Eliot's uncollected prose was consulted, in addition to the pieces republished in book-length collections. These items are most conveniently listed in Donald Gallup's essential *T. S. Eliot: A Bibliography*, Revised Edition (London: Faber & Faber, 1969). A new edition of this book is planned for 1988, and is likely to include a considerable number of pieces published or rediscovered since 1969. Some of these which have been of particular use in the present study are listed in the article by Eames and Cohn (see section 5 below). One piece not there listed, but important for Eliot's anthropological concerns is his 'Durkheim', *Saturday Westminster Gazette*, 19 Aug. 1916, p. 14. I am grateful to Dr Gallup for drawing this article to my notice.

4. BACKGROUND AND CRITICAL BOOKS

For reasons of space, this list is confined almost exclusively to important books dealing directly with T. S. Eliot. It does not include all books mentioned in the notes or in the catalogue of anthropological books given in Chapter III above. It does, however, include repeatedly cited works, for ease of reference. Where several works by one author are cited, these are arranged in chronological order, to facilitate cross-reference with the notes.

Ackroyd, Peter. *T. S. Eliot*. London: Hamish Hamilton, 1984.

Aiken, Conrad. *Ushant: An Essay*. 1952; rpt. London: W. H. Allen, 1963.

____ *Selected Letters of Conrad Aiken*, ed. Joseph Killorin. New Haven: Yale University Press, 1978.

Bergonzi, Bernard. *T. S. Eliot*. 2nd edn. Masters of World Literature. London: Macmillan, 1978.

Braybrooke, Neville, ed. *T. S. Eliot: A Symposium for His Seventieth Birthday*. New York: Farrar, Straus & Cudahy, 1958.

Browne, E. Martin. *The Making of T. S. Eliot's Plays*. Cambridge: Cambridge University Press, 1969.

Bush, Ronald. *T. S. Eliot: A Study in Character and Style*. New York: Oxford University Press, 1984.

Cornford, F. M. *The Origin of Attic Comedy*. London: Edward Arnold, 1914.

Costello, Harry T. *Josiah Royce's Seminar, 1913-1914: As Recorded in the Notebooks of Harry T. Costello*, ed. Grover Smith. New Brunswick, NJ: Rutgers University Press, 1963.

Durkheim, Emile. *The Elementary Forms of the Religious Life*, trans. Joseph Ward Swain. London: Allen & Unwin, 1915. 7th impression 1971.

____ *The Rules of Sociological Method and Selected Texts on Sociology and its Method*, ed. Steven Lukes, trans. W. D. Halls. London: Macmillan, 1982.

Frazer, J. G. *The Golden Bough: A Study in Magic and Religion*. 3rd edn., 12 vols. London: Macmillan, 1911-15.

Gardner, Helen. *The Art of T. S. Eliot*. 1949; rpt. London: Faber & Faber, 1968.

____ *T. S. Eliot and the English Poetic Tradition*. The University of Nottingham Byron Foundation Lecture 1965. Nottingham: The University, 1965.

____ *The Composition of 'Four Quartets'*. London: Faber & Faber, 1978.

Gordon, Lyndall. *Eliot's Early Years*. Oxford: Oxford University Press, 1977.

Gray, Piers. *T. S. Eliot's Intellectual and Poetic Development, 1909-1922*. Brighton: Harvester, 1982.

Hargreave, Nancy Duvall. *Landscape as Symbol in the Poetry of T. S. Eliot.* Jackson: University Press of Mississippi, 1978.

Harrison, Jane. *Themis: A Study of the Social Origins of Greek Religion.* Cambridge: Cambridge University Press, 1912.

Howarth, Herbert. *Notes on Some Figures Behind T. S. Eliot.* London: Chatto & Windus, 1965.

King, Irving. *The Development of Religion: A Study in Anthropology and Social Psychology.* New York: Macmillan, 1910.

Lewis, Wyndham. *The Letters of Wyndham Lewis*, ed. W. K. Rose. London: Methuen, 1963.

Lévy-Bruhl, Lucien. *How Natives Think* (trans. of *Les Fonctions mentales dans les sociétés inférieures*, tr. L. A. Clare). London: Allen & Unwin, 1926.

Litz, A. Walton, ed. *Eliot in his Time: Essays on the Occasion of the Fiftieth Anniversary of 'The Waste Land'.* Oxford: Oxford University Press, 1973.

Matthews, T. S. *Great Tom: Notes Towards the Definition of T. S. Eliot.* London: Weidenfeld and Nicolson, 1974.

Matthiessen, F. O. *The Achievement of T. S. Eliot: An Essay on the Nature of Poetry, with a Chapter on Eliot's Later Work by C. L. Barber.* 3rd edn. New York: Oxford University Press, 1958.

Moody, A. D. *Thomas Stearns Eliot, Poet.* 1st paperback edn. with additional appendix. Cambridge: Cambridge University Press, 1980.

Pound, Ezra. *Guide to Kulchur.* London: Peter Owen, 1938, rpt. 1966.

____ *The Selected Letters of Ezra Pound, 1907–1941*, ed. D. D. Paige. London: Faber & Faber, 1950.

Ricks, Beatrice. *T. S. Eliot: A Bibliography of Secondary Works.* Scarecrow Author Bibliographies, Number 45. Metuchen, NJ: Scarecrow Press, 1980.

Smith, Carol H. *T. S. Eliot's Dramatic Theory and Practice: From 'Sweeney Agonistes' to 'The Elder Statesman'.* Princeton: Princeton University Press, 1963.

Smith, Grover. See also under Harry T. Costello (1963).

____ *T. S. Eliot's Poetry and Plays, A Study in Sources and Meaning.* 2nd edn. Chicago: University of Chicago Press, 1974.

____ *The Waste Land.* Unwin Critical Library. London: George Allen & Unwin, 1983.

Soldo, John J. *The Tempering of T. S. Eliot.* Ann Arbor: UMI Research Press, 1983.

Tambimuttu and Richard March, ed. T. S. Eliot: A Symposium. New York: Tambimuttu & Mass, 1948.

Tate, Allen, ed. *T. S. Eliot: The Man and his Work.* London: Chatto & Windus, 1967.

Tylor, E. B. *Primitive Culture: Researches into the Development of Mythology, Philosophy, Religion, Language, Art, and Custom.* 2 vols., 3rd edn. London: John Murray, 1891.

Unger, Leonard. *Eliot's Compound Ghost.* University Park: Pennsylvania State University Press, 1981.

Vickery, John B. *The Literary Impact of 'The Golden Bough'.* Princeton: Princeton University Press, 1973.

Weston, Jessie L. *From Ritual to Romance.* Cambridge: Cambridge University Press, 1920.

Woolf, Virginia. *The Letters of Virginia Woolf,* ed. Nigel Nicolson and Joanne Trautmann. 6 vols. London: The Hogarth Press, 1975–80.

Wundt, William. *Elements of Folk Psychology: Outlines of a Psychological History of the Development of Mankind* (tr. Schaub). London: Allen & Unwin, New York: Macmillan, 1916.

5. ARTICLES

Again, this list is extremely selective. The criteria for inclusion, and the layout are the same as for the preceding section.

Beringause, A. F. 'Journey Through *The Waste Land'. South Atlantic Quarterly,* Jan. 1957, pp. 79–90.

Bluck, Robert. 'T. S. Eliot and "What the Thunder Said"'. *Notes and Queries,* Oct. 1977, pp. 450–1.

Clark, John A. 'On First Looking into Benson's *Fitzgerald'. South Atlantic Quarterly,* Apr. 1949, pp. 258–69.

Crawford, Robert. 'James Thomson and T. S. Eliot'. *Victorian Poetry,* Spring 1985, pp. 23–41.

——— 'Rudyard Kipling in *The Waste Land'. Essays in Criticism,* Jan. 1986, pp. 32–46.

Day, Robert A. 'The "City Man" in *The Waste Land*: The Geography of Reminiscence'. *PMLA,* June 1965, pp. 285–91.

Eames, Elizabeth R., and Alan M. Cohn. 'Some Early Reviews by T. S. Eliot (Addenda to Gallup)'. *Papers of the Bibliographical Society of America,* 1976, pp. 420–24.

Everett, Barbara. 'Eliot in and out of *The Waste Land'. Critical Quarterly,* Spring 1975, pp. 7–30.

Harmon, William. 'T. S. Eliot's Raids on the Inarticulate'. *PLMA,* 1976, pp. 450–9.

——— 'T. S. Eliot, Anthropologist and Primitive'. *American Anthropologist,* Dec. 1976, pp. 797–811.

Le Brun, Philip. 'T. S. Eliot and Henri Bergson'. *Review of English Studies,* 1967, pp. 149–61 and 274–86.

Lindsay, Maurice. 'John Davidson—The Man Forbid'. *Saltire Review*, Summer 1957, pp. 54–61.

Litz, A. Walton. 'That Strange Abstraction, "Nature": T. S. Eliot's Victorian Inheritance'. In *Nature and the Victorian Imagination*, ed. U. C. Knoepflmacher and G. B. Tennyson. Berkeley: University of California Press, 1977, pp. 470–88.

Materer, Timothy. 'Chantecler in "The Waste Land"'. *Notes and Queries*, Oct. 1977, p. 451.

Raine, Craig. 'Met Him Pikehoses: *The Waste Land* as a Buddhist Poem'. *Times Literary Supplement*, 4 May 1973, pp. 503–5.

Schuchard, Ronald. 'T. S. Eliot as an Extension Lecturer, 1916–1919'. *Review of English Studies*, 1974, pp. 163–73 and 292–304.

Soldo, John J. 'Eliot's Dantean Vision, and his Markings in his Copy of the *Divina Comedia*'. *Yeats Eliot Review*, vol. 7, no. 1 and 2, 1982, pp. 11–18.

Stead, William Force. 'Some Personal Impressions of T. S. Eliot'. *Alumnal Journal of Trinity College* [Washington], Winter 1965, pp. 59–66.

Stepanchev, Stephen. 'The Origin of J. Alfred Prufrock'. *Modern Language Notes*, June 1951, pp. 400–1.

6. RUNS OF PERIODICALS SCANNED

Criterion. Oct. 1922–Jan. 1939.
Dial. Jan. 1906–Sept. 1928.
Harvard Advocate. Oct. 1906–June 1914.
Harvard Crimson. Oct. 1906–June 1907.
Nation (New York). Jan. 1906–July 1914.
St. Louis Daily Globe-Democrat. 1 Jan. 1897–7 June 1897.

INDEX

Aborigines (Australian) 4, 19, 95-7,
    98, 122, 140, 142, 146, 179
Adams, Henry 77, 123
Adams, John 138
Aeschylus 160, 210
Aiken, Conrad 40, 44, 67, 68, 69, 71,
    79, 84 n., 99, 100, 101, 103
Alton, Illinois 12
Anesaki, Masaharu 30, 31 n.
Aristophanes 160, 163
Aristotelian Society 101
Aristotle 103
Arnold, Sir Edwin 31, 72
Arnold, Matthew 45, 61, 150, 153,
    158, 166, 197, 198, 200, 224,
    225
Athenaeum 98
Augustine, St 127, 128

Babbitt, Irving 61
Baedeker, *London* 11 n., 108, 117
Bantock, G. H. 224
Bastian, Adolf 191
Bateson, William 63
Baudelaire, Charles 11, 37, 44-5, 46,
    55, 57, 60, 82, 83, 84, 93,
    182-4
    'A une Madone' 81
    Intimate Journals 195
    'Les Sept Vieillards' 11, 45
    'Le Voyage' 181
Bédé, J. A. *see* Cailliet, E., and Bédé,
    J. A.
Bell, Clive 153
Bell, Bishop George 208
Benson, A. C. 121
Bergson, Henri 7, 8, 61, 63, 67, 93,
    199
    Creative Evolution 64, 67
Bible 25, 97, 140, 146, 158, 197, 203,
    205, 217
Blast 84
Blood, B. P. 72, 80
Boston 29, 30, 53, 73-4, 76, 81, 84,
    99, 123, 143, 196, 197
Bradlaugh, Charles 66

Bradley, F. H. 52, 81, 136, 181, 182,
    233
Brecht, Bertolt 207
Bristol, L. M. 89
Brooke, Brian 62
Browne, E. M. 210, 235
Browning, Elizabeth Barrett 36
Browning, Robert 159, 197
Brucke, R. M. 80
Bubu de Montparnasse (C. L.
    Philippe) 195
Buddhism 30, 31, 52, 71, 72, 82, 100,
    126, 127, 128, 139, 145, 149,
    186, 197, 220, 233
Burbank, Luther 65-6, 70
Bushnell, D. I. 14

Cahokia Mound Group 14
Cailliet, E., and Bédé, J. A. 199
Cambridge, Mass. 102 n., 195; *see also*
    Harvard
Campbell, Ian 43, 52
Canaletto 2, 66, 113, 114
Cape Ann 27, 33
Cats 4
Chambers, E. K. 138
Chapman, George 126, 182
Chapman (*Handbook of Birds*) 27
Chaucer, Geoffrey 128
Christianity 4, 31, 72, 73, 74, 79, 86,
    97, 116-20, 122-3, 126, 127,
    128, 131, 140, 141, 144, 145,
    149, 151, 154, 156, 158, 182-3,
    185-9, 192-3, 196, 198, 200-11,
    215-16, 218, 220, 223, 224, 228,
    230, 231, 233-5, 236; *see also*
    Church of England
Christian News-Letter 207
Church of England 19, 181, 189, 193,
    204
    Church of St Magnus Martyr 134,
        145, 147, 149, 192-3, 194
Codrington, R. H. 98, 150
Collingwood, R. G. 103
Conrad, Joseph 131, 159, 168, 169,
    170, 176, 192, 198, 221